Alexander Freier

Blockchain in the Energy Sector

An Advancing Technology to Tackle Global Climate Change?

With a foreword by Reimund Schwarze and Alejandro Bernhardt

This publication has been supported by Energiequelle GmbH, Zossen, Germany.

Alexander Freier

BLOCKCHAIN IN THE ENERGY SECTOR

An Advancing Technology to Tackle Global Climate Change?

With a foreword by Reimund Schwarze and Alejandro Bernhardt

Bibliografische Information der Deutschen Nationalbibliothek
Die Deutsche Nationalbibliothek verzeichnet diese Publikation in der Deutschen Nationalbibliografie; detaillierte bibliografische Daten sind im Internet über http://dnb.d-nb.de abrufbar.

Bibliographic information published by the Deutsche Nationalbibliothek
Die Deutsche Nationalbibliothek lists this publication in the Deutsche Nationalbibliografie; detailed bibliographic data are available in the Internet at http://dnb.d-nb.de.

Cover picture: Photo 118814407 © Blackboard373 | Dreamstime.com

ISBN-13: 978-3-8382-1717-8

Printed in the EU

For Laura & Gianluca –

The two most important people in my life.

Special thank you to
Juan Ignacio Ibañez & Michael Raschemann.

Foreword

The intersection of blockchain and energy is a topic that has attracted growing interest in recent years as the potential of this distributed ledger technology (DLT) to transform the energy sector becomes increasingly apparent.

However, despite its potential, the issue is relatively new to the academic debate, and there is still a significant gap in our understanding of how blockchain can be applied in the energy sector and the broader implication of this much-discussed innovation for global energy systems and society.

This book, *Blockchain in the Energy Sector*, and the Master's thesis I had the privilege of directing at the European University Viadrina Frankfurt/Oder in 2020 thus represents an important contribution to the literature on this topic. It combines international relations theory with management studies by using an innovative approach to address climate change as the most fundamental threat to the world's population and our ecosystems today. It provides a comprehensive overview of the potential implications of blockchain for the energy sector, its challenges, and how it could be applied to support the transition to a more sustainable energy system.

One of the key strengths of this book is its focus on international norms and how international technology transfer in the field of blockchain technology from a legal standpoint can be supported. This is a critical issue for the study of climate change, given the large unknowns about the potential of emerging technologies in general and those of blockchain, in particular, to drive down GHG emissions on a global basis.

So, while the book gives an interesting overview of how blockchain technology works, it also analyses two important use cases in the energy sector: The Brooklyn Microgrid and the WindNODE research project testing the viability of a renewable energy-based system in northeastern Germany. These use cases demonstrate blockchain technology's potential benefits and challenges in supporting the development of a more decentralized and efficient energy system. The lessons learned can and should be used to apply the technology to energy sectors in other parts of the world.

Overall, this book is an important first step in a research field that requires further exploration. It will interest scholars, practitioners, and policymakers concerning the potential of blockchain technology to 'green' our energy systems and support the transition to a more sustainable future.

Prof. Dr. Reimund Schwarze
Frankfurt/Oder, Germany May 2023

In our modern world, where complexity abounds and various fields intersect, we are witnessing a profound redefinition of disciplines. Notably, economics and management have expanded their horizons, embracing new dimensions of significance.

Gone are the days when policymakers and business leaders focused solely on maximizing macroeconomic gains, shareholder profitability, and customer satisfaction. A transformative trend has emerged on a global scale, gaining momentum, as both public and private sectors recognize the need for a different impact: acknowledging that economic optimization alone is insufficient and must be accompanied by social and environmental considerations. Consequently, sustainability has rightfully found its place and is reshaping the behavior of political and economic actors.

Such an approach, particularly in economics, calls for an interdisciplinary methodology integrating diverse knowledge strands. We can strive to maximize all the factors at play through seamless coordination simultaneously.

Scholarly work is of significant value within this complex landscape for several compelling reasons. At its core is the author, Alexander Freier, whose research presents a testament to his extensive and diverse academic background. Additionally, his extensive professional experience enhances this contribution, enabling him to adopt a multifaceted approach from various angles. Furthermore, his global experiences, studying and working in different cultural contexts, provide him with a comprehensive lens to understand the multiple layers inherent in our subject matter.

Through the chosen approach, the author aptly underscores the paramount significance of fostering renewable energy resources to bolster global sustainability efforts. Furthermore, introducing blockchain as a nascent technology broadens the reader's understanding of potential avenues for innovation. The convergence of these two critical subjects lays the foundation for a compelling case, necessitating a reevaluation of business models that prioritize environmental responsibility and foster social inclusivity. This intersection promises to yield transformative possibilities whereby greener and more socially conscious frameworks can be envisioned and actualized. For these reasons, I wholeheartedly declare this book invaluable to addressing complex and consequential issues. Its ultimate goal is to advance the collective well-being of humanity, both in the present and future.

Finally, personally, I derive immense satisfaction from witnessing my former student and esteemed friend, Alexander, dedicating himself wholeheartedly to these pressing concerns. I firmly believe that he possesses the essential qualities and professional values necessary to tackle these challenges with the utmost effectiveness.

Alejandro Bernhardt, PhD.
Córdoba, Argentina May 2023

Abstract

Renewable energy has become a key research focus in global environmental governance and is regarded as a principal source in the struggle against global climate change. Although fossil fuel consumption has long been identified as the prime contributor to growing global greenhouse gas emissions, countries pursuing energy transitions toward the large-scale insertion of renewable sources continue to experience constant emission increases, as well. In particular, the imbalance of renewable energy demand and supply due to the volatile generation of solar and wind energy has been named the main reason for energy losses and the lack of final renewable energy consumption.

Digitalizing energy transitions and implementing new technical innovations to enhance the efficient usage of renewable energy inspire researchers in various academic fields. Blockchain, a DLT using cryptography, has recently gained prominence as an integral component within the increasingly digitalized infrastructure surrounding the energy sector.

Against this background, the main argument brought forward in this book is that applying blockchain technology can be a viable option to both enhance the efficiency and balance of renewable energy generation and consumption and to reduce GHG emissions if the following three components subsequently are met: technological advancements, an adequate international normative framework, and a general trust on behalf of key market actors to promote "blockchain energy" on a global scale.

To highlight the concerted global efforts through which climate change can be fought exclusively and to prove the hypothesis, this book will theoretically be embedded within international relations approaches. Departing from the world's climate change-energy nexus analysis, a legal analysis to explore possible "bottom-up" approaches for implementing and transferring technological innovation deriving from the international climate contracts shall be conducted. Based on this analysis, the technological infrastructure surrounding blockchain energy shall be explored and applied to the case studies of the Brooklyn Microgrid in the United States and the WindNODE trading platform in Germany.

Table of Contents

List of Figures

List of Abbreviations

ADS	Active Distributed System
AND	Active Distributed Networks
BMWi	Federal Ministry for Economic Affairs and Energy of Germany
BNetzA	Federal Network Agency of Germany
CHP	Combined Heat and Power
CDM	Clean Development Mechanism
COP	Conference of the Parties
DA	Distribution Automation
DER	Distributed Energy Resources
DLT	Distributed Ledger Technology
DSO	Distribution System Operator
DR	Demand Response
EE	Energy Efficiency
EIA	International Energy Agency
EPRS	European Parliamentary Research Service
ET	Emission Trading
EU ETS	EU Emission Trading System
GHG	Greenhouse Gas
GIZ	German Agency for International Cooperation GmbH
IAEA	International Atomic Energy Agency
IPCC	International Panel on Climate Change
IR	International Relations
JI	Joint Implementation
LEM	Localized Energy Market
LV	Low Voltage
MCP	Market Clearing Price
MW	Megawatt
OECD	Organization for Economic Cooperation and Development
OPEC	Organization of the Petroleum Exporting Countries
PoW	Proof-of-Work
PoS	Proof-of-Stake

RE	Renewable Energy
RES	Renewable Energy Source
T&D	Transmission and Distribution
TSO	Transmission System Operator
UNFCCC	United Nations Framework Convention on Climate Change

1. Introduction and Hypothesis

1.1. Introduction

Global climate change has become one of the most important threats to human security today[1] (Vivekananda, 2022; Dumaine & Mintzer, 2015; Adger *et al.*, 2014; Dalby, 2013). The global dimension of climate change and its diverse local manifestations on people's lives and livelihoods requires identifying the central aspects contributing to this challenge. The key link between climate change and greenhouse gases such as CO_2 is primarily found in the extraction and burning of non-renewable, fossil fuel-based energy resources (Soeder, 2021; Steen, 2001). Given its portability and comparatively affordable prices, global petroleum and gas supplies represent the basis for most of the world's societies and industries. These dependencies have created "global energy dilemmas," which led to the increasing production and consumption of fossil fuel resources despite the large-scale expansion of RE sources on a worldwide basis (Bradshaw, 2010). Although large investments in sustainable technologies and efforts to increase EE continue to take place, more than 85 percent of the primary energy consumed stems from extracted fossil fuels. This accounts for 56.6 percent of all

[1] According to Resolution 66/290 of the United Nations (UN) adopted on 10.09.2012, human security refers in its first three articles to: "(a) The right of people to live in freedom and dignity, free from poverty and despair. All individuals, in particular vulnerable people, are entitled to freedom from fear and freedom from want, with an equal opportunity to enjoy all their rights and fully develop their human potential; (b) Human security calls for people-centred, comprehensive, context-specific and prevention-oriented responses that strengthen the protection and empowerment of all people and all communities; (c) Human security recognizes the interlinkages between peace, development and human rights, and equally considers civil, political, economic, social and cultural rights" (United Nations General Assembly, 2012, p.1). In order to achieve these objectives, the report highlighted the economic, health, personal, political, food, environmental and community dimensions as essential parts of to the agenda (Gómes & Gasper, 2013, p.2). Drawing on these ideas, Mason argues that: "What can be readily acknowledged is that there are diverse trajectories of climate-related influence on human lives and livelihoods, though it is the severe stress on vulnerable peoples attributed to present and future climate change that has justified its 'human securitization'" (Mason, 2013, p.382)

global anthropogenic greenhouse gas emissions (Moomaw *et al.*, 2012, p.164).

Given these developments, breaking through the vicious cycle of fossil fuel-based energy security on the one hand and the necessity to reduce global CO_2 emissions to keep temperature increases well below the two-degree margin as envisaged by the 2015 Paris Accord on Global Climate Change on the other, additional concepts and methods of energy consumption need to be developed, urgently. Consequently, CO_2 emissions can only be substantially reduced if the correlation between traditional pattern of fossil fuel energy consumption and CO_2 as the main contributing factor to climate change is being acknowledged (Akhmat *et al.*, 2014). As the above-described developments indicate however, it remains a daunting task to switch to alternative resources to reduce GHG emissions and "green consumption pattern[s]" have shown to not necessarily lead to a significant reduction in CO_2 emissions (Alfredsson, 2004). Instead, there is a strong probability that energy consumption will furthermore increase as a result of the global impacts of climate change (van Ruijven, 2019).

One major component in the global efforts to reduce CO_2 emissions is represented by the massive expansion of RE resources (Kung & McCarl, 2018). Such a transformation is important as fossil fuel-based economies have been shown to enhance CO_2 emissions further, while renewable energy-based ones contribute to their short- and long-term reductions (Ali & Seraj, 2022). However, the difficulties displayed in countries undergoing energy transitions like Germany call the attainability of the combined climate and energy objectives into question. Having started with its ambitious goals to phase out both nuclear power plants and fossil fuels after the Fukushima nuclear accident and still achieve climate neutrality by 2045 (after the legally enforced adaptations to the National Climate Act in 2021) represents a tremendous challenge (Federal Constitutional Court, 2021; German Federal Government, 2021). Although the country can (theoretically) replace all of its closed-down nuclear power plants with other power resources, the electricity stemming from atomic energy between 2010 and 2019 has mainly been replaced by coal-fired production and net electricity imports.

This holds even when investments in renewables are taken into account (Jarvis *et al.*, 2022).

Contemporary difficulties for nearly all major energy transitions on a global scale continue to result from the *volatility* of renewable power generation—primarily from wind and solar. Currently, available storage mechanisms do not possess the capacity to comprehensively make sufficient electricity generated from renewables available at all times (Sinn, 2017, pp.131–132). The roll-out of green hydrogen as another much-discussed option is expected to be too limited in the medium term (Bard *et al.*, 2022), entails long-term investment subsidy lock-in risks for hydrogen technology (Menner & Reichert, 2020), and is not yet economically competitive if compared to fossil fuel-based forms of hydrogen (Agora & Guidehouse, 2021, p.11).[2]

This background of limited success when relating RE transitions to reducing CO_2 and other GHG emissions increases the rethinking of energy generation, distribution, and trade based on digital innovations. The European Green Deal, whose main objective is to make the EU a carbon-neutral continent by 2050, perceives digitalization as central to the intended green transformation (Fetting, 2020, p.13). To achieve the objectives of the Green Deal, The European Commission furthermore called the digital and sustainable transformation as the key to achieve energy independence and to successfully carry out the RE acceleration plan RePowerEU (European Commission, 2022).

The Paris Accord recognizes the current difficulties resulting from the diverse and differing strategies to limit GHG emissions by promoting cooperative approaches broadly referred to as "internationally transferred mitigation outcomes" (ITMO) to enable participating countries to reach their Intended Nationally Determined Contributions (INDC). It is very problematic for multiple national jurisdictions and differing monitoring schemes, which call for

---

2 As of 2021, the cost of renewable hydrogen ranged between €3.40 and €6.60/kg and is mainly determined by the expenses for renewable electricity, the electrolyzers' capacity factor and the system costs of the later. Renewable hydrogen was therefore in average €3/kg more expensive than hydrogen stemming from fossil fuels (Agora & Guidehouse, 2021, p.11).

digital monitoring, reporting, and verification of carbon offsets (World Bank, 2022). Consequently, Article 6 of the Paris Agreement addresses CO_2 emissions by welcoming innovation processes and private investments into low-emission technologies (World Bank, 2018, p.4). Additionally, the Paris Agreement calls for "a 'bottom-up' country-driven implementation process" (Dzebo *et al.*, 2019, p.4), which takes different social and economic actors within distinct local settings into account, instead of relying exclusively on "top-down" international institutions-based approaches (Banda 2018; Zaman, 2018). Within the context of this book, "bottom-up" is referred to as "a model that represents a system by looking at its detailed underlying parts. Compared to so-called top-down models, which focus on economic interlinkages, bottom-up models of energy use and emissions can provide greater resolution with regards to sectors or mitigation technologies" (UNEP, 2017, p.vii).

The Paris Agreement takes this growing complexity into account and urges multilevel approaches to go about cutting emissions. Blockchain technology can play a central role if applied correctly to distinct climate markets.

Due to these developments, blockchain technology has gained prominence as a possible digital innovation within an envisioned sustainable infrastructure contributing to reducing GHG emissions (OECD, 2019).[3] Emphasized as one possible digitalization strategy by national governments, research institutions, international organizations, and private start-up companies, the wide array of application options (ranging from inter-company interactions, taxing, management of supply chains, peer-to-peer trading, demand-side response, carbon-offset trading, etc.) have made blockchain technology a focal point for research and investments. While only having a market share of around $1.2 billion in 2018, this amount is forecasted to rise to $23.3 billion in 2023. And after the financial sector, with Bitcoin trading as its major component, the energy sector

---

[3] Although secure blocks of data chains have been known since the early 1990s, the implementation of blockchain technology for production purposes only came about with the publishing of the White Paper proposing peer-to-peer electronic currency exchange, which is widely known as Bitcoin (Nakamoto, 2009).

will be the second biggest beneficiary of these investments (GIZ, 2019a, p.3).[4]

At its core, blockchain is a transaction management tool based on digital data sharing. Its cryptographic methodology allows for accurate verification of all past and present transactions. These transactions are conducted on a so-called peer-to-peer network basis, representing the basis for this digital accounting system's trustworthiness (Hawlitschek *et al.*, 2018; Wei *et al.*, 2019). So, in general terms, blockchain's novelty lies within the gradual substitution of the "middle-man" and, therefore, the intermediate institution to manage the interaction between two or more parties.

Although unlikely to entirely sidestep central managing authorities in the near future (as in areas such as the interoperability of climate incentive schemes and RE), rethinking governance frameworks oriented toward international problem-solving calls for the large-scale inclusion of innovations such as blockchain technology, artificial intelligence, sensor networks, and FinTech applications (GIZ 2019b, p.85). In other words, while the reliance on third parties for verification, security, and privacy purposes, especially for digital assets and properties, represents a major source of instability and is prone to risks of manipulation, the simultaneousness of distributed consensus and anonymity of blockchain displays key

[4] Updating this book in the post-Covid19 pandemic era, which brought large parts of the global economy to a unique standstill, new and digitalized working trends expanded on an unprecedented scale to guarantee business continuity. Online-based communication altered the world's working culture (Savić & Dobrijević, 2022; Błaszczyk *et al.*, 2022; Madhumathi *et al.*, 2021). Thus, using blockchain-based technology became a largely discussed issue and increasingly applied method (Shah *et al.*, 2022; Hilal *et al.*, 2022; Botene *et al.*, 2021). The complexity and interconnectedness of today's world economy and the rapid speed at which the world society turned to digital innovations to adapt to these challenges have furthermore elevated the importance of blockchain technology (Kalla et al., 2020). Unsurprisingly, therefore, the Covid-19 pandemic actually contributed to the creation of a higher level of blockchain stability. As shown by ABI Research, blockchain revenues fell by approximately 35%, equaling a loss of $2.8 bn between 2018 and 2020. And, while the latter continues to be a volatile process and is mainly attributed to price drops and the subsequent disappearance of around 2000 cryptocurrencies, the urgency for increased transparency and tamper-proof interactions demand the ongoing digital transformation especially in times of these above-described tendencies (ABI Research, 2020).

benefits in terms of transparency (Crosby *et al.*, 2015; GIZ, 2019a). Consequently, this very openness and transparency, combined with the (potentially future cost and) risk reduction benefits, have made blockchain interesting for investors across various economic sectors. Although a methodologically and theoretically little-explored topic, blockchain's disruptive character will likely promote a reconfiguration of today's economic, political, legal, and cultural space (Frizzo-Barker *et al.*, 2020; Swan, 2015).

With regard to this research, which empirically focuses on blockchain-based RE trading in the United States and Germany, this book contains inasmuch a development perspective as it embeds the bottom-up approaches into an international relations theory framework by showing how energy distribution can *potentially* be made more efficient and socially inclusive by applying blockchain technology. Because more than 789 million people did not have access to steady electricity provision in 2018, investing in blockchain-based energy trading and distribution technology can be an option to reduce CO_2 emissions and mitigate energy poverty. And even though the number has steadily decreased (from 1.2 billion in 2010), the Energy Progress Report 2020 shows that the total final energy consumption of renewables increased only by 1.0 percent (to 17.3 percent) between 2010 and 2017 (World Bank, 2020, p.1). Goal 7 of the 2030 Agenda for Sustainable Development proclaims the target objective of "a world where human habitats are safe, resilient, and sustainable and where there is universal access to affordable, reliable and sustainable energy" (United Nations, 2016, p.5). At the current pace of investments into the RE infrastructures on a global scale, however, increasing the initially proclaimed level of human security concerning clean energy access options is unlikely to be reached.

Considering, on the other hand, the learning curve in the innovation of RE and the immense price drop for the purchase of solar power modules of 80 percent between 2008 and 2012 alone (Jäger-Waldau, 2019, p.46), it becomes clearer that the technical and financial opportunities to meet the sustainable development goals do exist.

Against this backdrop, this book will explore the potential of blockchain-based RE distribution by analyzing the Brooklyn Microgrid in New York and the Fraunhofer Institutes' research regarding the WindNODE flexibility trading platform in northeastern Germany. This book, therefore, picks up on the above-described technological developments in the field of blockchain technology, looks at their potential to balance the demand and supply of RE, and, by doing so, explores "blockchain energy" for its potential viability to reduce GHG emissions.

1.2. Hypothesis

The underlying hypothesis of this research study posits that applying blockchain technology can be a viable option to both enhance the efficiency and balancing of RE generation and consumption and to reduce GHG emissions if the following three components subsequently are met: technological advancements, an adequate international normative framework, and general trust on behalf of key market actors to promote blockchain in the energy sector on a global scale.

2. Methodology and Justification of the Chosen Research Topic

Approaching the chosen topic represents both theoretically and empirically challenging tasks. Departing from the idea of continuously growing global GHG emissions even after the large-scale implementation of RE resources, exploring how blockchain technology as a central component to the ongoing digitalization process of the world's energy transitions can contribute to a balancing of RE supply and demand and a more sustainable form of energy consumption is a complex endeavor.

As outlined in the introduction, this book explores bottom-up initiatives and opportunities to highlight the importance of blockchain technology in decentralized RE projects adapted to distinct local and regional settings to enhance green energy consumption efficiently. However, a combination of top-down and bottom-up approaches was chosen to embed these bottom-up approaches within an encompassing global and contextual analysis surrounding the emergence of blockchain technology in the energy sector. As was pointed out, the first term refers to an aggregated and the latter to a disaggregated model. So, while a macroeconomic perspective characterizes the first approach, the latter focuses on the specifics of a given technical innovation in the energy sector (Böhringer & Rutherford, 2007).

Going about the research in such a way, it needs first to compose a theoretical frame allowing for the analysis of the CO_2—climate change—energy nexus and the normative international framework to promote the transfer of technical innovations such as blockchain as a possible climate technology in the energy sector. Such an analysis is, therefore, based on a macro perspective. On the other hand, technological innovations (although usually consisting of "global" intellectual knowledge) are built at distinct locations and need to be applied within distinct local infrastructural settings. Consequently, international policy frameworks can only unfold their intended impact if international promotion schemes acknowledge the characteristics and capabilities of new technical innovations. Here, the bottom-up analysis is of importance.

Following this, a selection and explanation of international relations theories will be conducted, which allow for the explanation and analysis of global climate governance and the use of technological innovations like blockchain as a climate technology from both a top-down and bottom-up perspective. In this book, Global Governance, Regime Theory (Liberal Institutionalism), Policy Learning and Diffusion, and Green Theory have been chosen.

The link between CO_2 emissions, climate change, and the energy sector is then outlined to highlight their strong correlation. This has mainly been done based on scientific literature and statistical analysis.

In the following step, a legal analysis of the normative framework of the climate contracts of the United Nations will be conducted. This allows, first of all, for exploration as to which IR theory best fits to explain the status quo of these contracts. However, it also allows for detecting bottom-up strategies and using technological innovations such as climate technologies. To do so, climate contracts, as displayed on the website of the UNFCCC and in selected scientific literature, are primarily studied and analyzed.

Thereafter, so-called smart technologies and their capacity to balance demand and supply and emerging consumer behavior trends such as "prosumption" to better take advantage of these technological advancements and fully unfold their potential within the energy sector will be analyzed. Furthermore, necessary elements for applying blockchain to the energy sector, such as smart grids, DERs, smart contracts, and possible limitations of blockchain, shall be explored and described. To do so, a state-of-the-art literature review is being conducted.

In the following step, the potential of blockchain-based microgrids is being researched. After analyzing state-of-the-art technology mainly based on accessible (engineering) papers, this book will empirically focus on the Brooklyn Microgrid (BMG). The research analyzes existing scientific papers, published newspaper articles, and additional qualitative and quantitative information.

Second, the WindNode blockchain-based trading platform in Germany will be analyzed as the second use case. This research could count on one expert interview with the technical lead of the blockchain prototype at Fraunhofer FOKUS and an expert interview with a UCL's Centre of Blockchain Technologies member. These interviews were necessary as scientific information on this particular research project was scarce in 2020. The information provided shall be embedded into a wider context regarding the ongoing digitalization of Germany's energy transition.

Finally, to carry out this work in a methodologically valid and coherent way in terms of text selection, text legal interpretation as well in conducting expert interviews while maintaining the correct focus on epistemological guiding principles of objectivity and generalizability, this work is based on the methodological principles outlined by Opp (2002) and Schnell *et al.* (1999).

This book is based on the author's Master's thesis written, turned in, and defended at the European University Viadrina Frankfurt/Oder (Germany) and later at the Catholic University of Córdoba (Argentina) as part of a double-degree program. In accordance with the German publishing house, parts of the thesis have been pre-published on the open-source website for scientific articles SSRN to receive additional academic feedback for this book (see: Freier, 2022a; Freier, 2022b).

On a personal note, the main challenge regarding the research for this book has been the parallel onset of the global Covid-19 pandemic and the lockdown since March 2020, which continued for the best part of 2021. The latter not only affected applying new technologies to the energy sector, but the options to conduct active infield research had therefore been extremely limited. As a result, the chosen use cases, and the applied methodology for their analysis result from the conditions under which the pandemic played out.

3. Theoretical Approaches

3.1. Introduction

The study and research field of environmental governance in general and climate change, in particular, are facing major challenges in theory formation across different academic disciplines. In the words of Young, "we have made little progress as a community of scientists and practitioners toward the development of a unified theory of environmental governance" (Young, 2005, p.182) at a time in history when the necessity to act on a coordinated international basis to reduce GHG emissions became a visibly urgent task. Instead, a multitude of different theories approaching the topic from different angles and with different perspectives continue to dominate the academic sphere. Given the complex nature of environmental governance, only distinct elements of the topic can be theoretically and methodologically approached, which vary due to their diverging ontological and epistemological assumptions (Beunen et al., 2022, p.2).

In international relations, in particular, environmental governance approaches have mainly borrowed from other theories (as in the case of institutionalist strands from Keohane, 1984; Keohane & Victor, 2011). In a discipline largely caught in a vicious cycle of peace, security, and order-building analysis, issues such as greenhouse gas emissions and the diverse local manifestations of climate change have mainly been regarded as "threat multipliers" of existing conflict potentials (Vogler, 2018).[5] Consequently, a variety of top-down and bottom-up approaches coexist, which are trying to explain varying governmental behavior according to a norm-based constructivist (Bueno, 2019; Demeritt, 2006; Stehr & von Storch, 1995) or rational choice-based climate complex regime theories (Keohane & Victor, 2011; Vogler, 2018; Brunner, 2001).

[5] The UN Security Council continues to highlight that "the relationship between climate-related risks and conflict is complex and often intersects with political, social, economic and demographic factors" (United Nations, 2019).

Resulting from these observations in theory building, definitions of environmental governance remain rather broad, and the term shall equally broadly be perceived as a:

> "(...) structure that enables society to manage the environment, just as a conductor unifies and adjusts the overall performance of an orchestra. It is thus useful to place the theory of "governance" within the domain of environmental economics as it searches for an analysis that will provide a concrete policy structure for the environmentally sensitive conservation society" (Yoshida, 2012, p.77).

Given these limits to tackling environmental governance due to the complex interplay of overlapping factors leading to global GHG emissions and environmental degradation, a variety of international relations theories shall be presented and later evaluated for their usefulness in explaining the usage of emerging technologies like blockchain as a potential climate technology contributing to more efficient use of (renewable) energy.

3.2. Global Governance

The potential threat of climate change has and continues to be not as tangible, and many of the resulting economic impacts remain unquantifiable (Rising *et al.*, 2022). Therefore, global governance theory calls for a rethinking of the logic of international politics, especially in times of growing risk potential not only stemming from security-related issues and shifts away from a pure government perspective by including diverse non-governmental actors and civil society networks, their demands for a higher level of transparency, participation, justice, and equity, while also paying attention to topics such as the emergence of new technologies and their social implications (Hira & Cohn, 2003, p.11). Thus, climate change represents one of the key global challenges only to be successfully manageable on a concerted basis and at different societal levels. According to Mayntz, "ecological globalization" includes all the negative environmental externalizations whose effects cause ecological manifestations beyond national borders. The depletion of the world's ozone layer, global warming, and the loss of flora and fauna demand collective efforts to address these challenges effectively

(Mayntz, 2002, p.1). From these observations, a definition of global governance arises, which proclaims that:

> "(...) international governmental and non-governmental organizations make up what is generally called global governance. Governance is by definition about collective problem-solving, not about dominance for its own sake. Both international governmental and non-governmental organizations are involved in processes of collective problem solving; the problems they are supposed to address are put down in their statutes and their stated missions. This holds for supranational institutions like the United Nations, for international governmental organizations like the International Labor Office, the World Bank, and the World Trade Organization, and for international regimes dealing with problems of the global ecology and human rights issues. International public interest organizations similarly address a range of humanitarian and ecological problems. Even international business associations do not only function as pressure groups for economic interests; in addition to their involvement in international decision processes, organizations such as the OECD or the International Chamber of Commerce also perform some regulatory functions" (Mayntz, 2002, p.2).

Such a definition of global governance stands in the liberal tradition of international relations and gained prominence from the beginning of the 1990s onward as a possible solution to tackle the multitude of growing global challenges. Rosenau and Czempiel (1992) called for including a variety of (non-state) actors in a newly emerging global governance order, which has increasingly been ascribed legitimacy due to their expertise and general recognition as knowledge actors within a broader societal context. In other words, the increasing economic interdependence demands new and collective approaches by different legitimate local and global actors to adapt to growing complexities and uncertainties. As pointed out by Cadman, legitimacy can thereby be input- and output-oriented. While the first refers to the acceptance of rules by those being governed, the latter focuses on the efficiency of rules for good governance. In climate governance, in particular, both forms of legitimacy usually depend on each other, and profound interaction between different interest groups is necessary to meet and align local community necessities and interests with the content of discussions and policies on the international level (Cadman, 2013, p.8).

Although there are various similarities between global governance and regime theory, especially regarding legitimacy, a key difference can be detected since "global governance is not obsessed with order and cooperation among states. In regard to equity, the emergence of global civil society, and the need for greater accountability of global regimes, global governance is rather prospective and normative" (Hira & Cohn, 2003, p.11).

3.3. Regime Theory

Keohane (1984) developed a theory based on the idea that anarchy—a key ordering principle of the international system proposed by theories of realism—can be overcome because growing interdependence forces rational actors into cooperation. Consequently, regime theories have gained importance as problem-solving approaches. According to Krasner, regimes can be defined as:

> "(…) sets of explicit principles, norms, rules, and decision-making procedures around which actors' expectations converge in a given area of international relations. Principles are beliefs of fact, causation, and rectitude. Norms are standards of behavior defined in terms of rights and obligations. Rules are specific prescriptions or proscriptions for action. Decision-making procedures are prevailing practices for making and implementing collective choice" (Krasner, 1983, p.2).

Regimes are created faster if the participating parties share common interests in creating them and if the threats of not finding a consensus on how to act based on the four regime elements shall cause negative consequences for the participant nations (Keohane, 1982). Based on this rational choice logic, defense and missile control agreements were created during the cold war. In the field of petroleum, 'energy security regimes'[6] such as OPEC as well as the International Energy Agency (IEA) were built.

---

6 According to Esakova (2012), an "energy security regime can be defined as governing arrangements that affect relationships of interdependence in the field of energy security. (…) The field of energy security is characterized by a high level of interdependence between energy markets, energy exporters and importers, etc. The cornerstones of an energy security regime are various transactions between actors involving the trade of energy" (Esakova, 2012, p.78).

7 As the approval of the so-called EU taxonomy has shown, however, this continues to be a daunting task. This taxonomy classifies certain gas and nuclear energy activities as sustainable (European Commission, 2023d).

Due to the complexities laid out in the previous sections, defining and specifying a regime within the field of climate change represents a much more diverse and complex endeavor. On the one hand, geographic, cultural, legal, technological, and economic diversities between countries and regions and the multiple factors contributing to GHG emissions make it difficult to find a common political ground to create a stable and efficient climate change regime. Given these challenges of finding political consensus and avoiding problems such as free-riding, the incentives for national governments to act within these highly specified regimes remain limited.

Therefore, in the aftermath of the failed Copenhagen Accord of 2009 and the Post-Kyoto-Process, Keohane and Victor coined the term "regime complex" to describe lightly interconnected regimes consisting of "functional, strategic, and organizational components" (Keohane & Victor, 2011, p.14). By doing so, they highlight the fact that regime complexes can display a higher level of flexibility and adaptability despite the lesser extent to which nations are bound to perform within them. On the contrary, however, they also point to the fact that regime complexes can lead to higher confusion and a lesser willingness to participate in consensus-building on various issues. This is likely to slow down concerted efforts in climate change-related areas. The following regime complex elements need to be fulfilled to enhance the chances of their long-term stability. First, Keohane and Victor refer to the fact that a regime or regime complex needs to be compatible and mutually reinforcing. Even if states can agree upon creating a regime, the individual procedures to reduce GHG emissions must be aligned and may not contradict each other. If this can be assured, clear rules for accountability, measuring effectiveness, determinacy, sustainability, and epistemic quality need to be established (Keohane & Victor, 2011, pp.19–20).

This regime's complex logic outlined above even applies after enacting the 2015 Paris Accord. Alter and Raustiala point out that:

> "(...) driven in part by the increasing dominance of game theory, it had become the practice in the international relations literature to theorize about cooperation largely in terms of states creating discrete regimes to govern discrete problems. Although this approach rendered international

> cooperation more analytically tractable, the dense and growing array of existing agreements and institutions means that it is increasingly inaccurate and distortive to imagine cooperation as if actors are proceeding on a blank slate. Instead, many scholars recognize that it is essential to focus on how the overlapping and sequential nature of international commitments in itself shapes the politics of cooperation, the interpretation of agreements, and the decisions of actors operating within and around regime complexes" (Alter & Raustiala, 2018, p.330).

So, although the Paris Accord as such is an important step forward in terms of creating a common framework for current and future cooperation in global climate policy, Keohane and Oppenheimer (2016) argue that the COP 21 Accord is by no means likely to maintain global warming below the two-degree level. The Paris Accord marked, at best, a kick-off of global climate politics by loosely defining (INDCs) and by referring to possible activities of domestically and internationally organized economic and social groups. Continuing in this line of argumentation, the authors point out that:

> "What is clear is that whether negotiations lead to substantial emissions cuts will not depend chiefly on the text of the Paris Agreement. It will depend much more on domestic and transnational politics within and between the OECD countries and the BRICs. This is to say that by itself the Paris Agreement accomplishes little— but it opens what was a locked door. That door is now a little bit ajar -pushing hard could carry us through it to a better outcome, but nothing will be accomplished at the international negotiation level alone. There will have to be pressure within the OECD countries for vigorous emissions action by wealthy states and for financial support for effective action in poorer countries. And that pressure will have to entail willingness to pay" (Keohane & Oppenheimer, 2016, p.150).

Institutions as regimes—and that might be perceived as a key benefit of the Paris Accord—incorporate certain incentive structures, which then have to be used by political and economic actors when implementing changes to national legislation and subsequent investments into low-carbon technologies. Policy, economics, knowledge, capacity, reputation, and socioeconomic incentives are among those stimulations supporting the introduction of changes, especially in climate-related decision-making (Rai & Nash, 2017, p.147).

These explanations highlight that political and economic engagement by influential domestic and international pressure groups largely depends on opportunities for long-term investments in climate technologies. With regard to the latter, the sources of possible incentives can be manifold. In other words, interest groups in the RE sector can gain ground if their investments' economic and political incentives emerge and allow for a possible long-term disruption of an economic infrastructure based on fossil fuel usage. Additionally, traditional energy producers such as the nuclear power and the petroleum sector can only be contested by renewables if the economic incentive structures favorable to their implementation come with a growing societal consensus and a political awareness that energy consumption patterns must be urgently altered.[7]

Although important in understanding the global environmental policies emerging around the negotiations of CO_2 emission reductions, regime theory, in its most classical perception, is—as outlined above—very much embedded within a state-centric worldview. However, this top-down approach has shown to be a difficult analytical tool when considering the cultural, economic, and socioeconomic diversities of countries, regions, and localities. The Kyoto Protocol of 1997, an international treaty (extended by the United Nations Treaty on Climate Change) obligating participating nations to reduce GHG concentrations, has been linked to the traditional top-down logic and ultimately failed to display significant emissions cuts. According to Keohane and Victor, the failure of the Kyoto regime resulted from a lack of obligations and incentives to commit to the GHG reduction targets and the dysfunctionality of international organizations (Keohane & Victor, 2015). However, the protocol has also been described as a:

> "(...) policy shift towards sustainability energy production - a shift that is likely to have significant impact on technology development and competitiveness assessments. Thus, although a modest beginning, the protocol's emission reduction regime has the potential to generate a dynamic towards progressively greater participation and deeper reductions. Less obvious to all but close observers may be another set of reasons for the significance of

[7] As the approval of the so-called EU taxonomy has shown, however, this continues to be a daunting task. This taxonomy classifies certain gas and nuclear energy activities as sustainable (European Commission, 2023d).

> the Kyoto Protocol, rooted in fact that it has taken negotiators into largely uncharted territory. The regime that has emerged offers international lawyers and policymakers a living laboratory for testing and refining new approaches to global environmental problem solving" (Brunnée 2003, pp.255–256).[8]

As these insights on the growing complexity of environmental governance illustrate, thinking in terms of bottom-up and, therefore, more decentralized terms resulted from regime negotiations. "Climate justice" also called for recognizing injustices regarding the materialization of CO_2 emissions.[9] Islam and Winkel (2017, pp.4–7) argue that climate change and inequality have created a vicious cycle that becomes further perpetuated through climate hazards. So, the aggravating effect of climate change on inequality made a shift from the prime focus on *mitigation* toward the *adaptation* to climate change necessary. The latter also requires enhanced bottom-up assessments of climate change-related risks especially when drawing attention to vulnerable environments.

Consequently, analyzing vulnerabilities has to be at the center of policymaking (Conway *et al.*, 2019). This does, however, not imply that top-down and bottom-up approaches are mutually exclusive. They rather complement each other and take part in a multi-level governance strategy. Given that climate change is a largely ethics and justice-related topic, both forms of governance need to be integrated to efficiently respond to the multifaceted socioeconomic, cultural, and legal challenges to reduce global GHG

---

8 So, while the Kyoto Protocol has been an important step forward in terms of promoting and generating global consciousness regarding the necessities to jointly combat man-made CO_2 emissions and a first step toward the creation of a climate change regime, the inefficiencies of top-down approaches to consider the complexities on issues like climate change became all too obvious. The failure to make some of the world's most polluting nations commit to GHG emissions reductions and/or to consent to a follow-up agreement illustrates the difficulties of hierarchical approaches while calling for an institutional rethinking of climate change policies (Carlarne, 2012).

9 While northern industrialized countries are the main contributors to GHG emissions, poverty-stricken southern nations are among those disproportionally suffering from the effects of climate change. And despite these unequal global impacts of climate change, the means to respond by poorer countries are much more limited and have diverse socioeconomic effects (Islam & Winkel, 2017, pp.4–7).

emissions (Dirix *et al.*, 2013). The UNFCCC recognized these difficulties and the need for more flexibility in international decision-making by allowing each state to develop its own suggestions (INDCs) regarding reducing global emissions. The bottom-up logic has been considered, allowing for more suitable and flexible local and regional contributions. These efforts should be accompanied by sufficient "inter alia, mitigation, adaptation, finance, technology development, and transfer, capacity building, and transparency of action and support" (UNFCCC, 2013, p.6).

3.4. Policy Transfer and Policy Diffusion

Given the difficulties in conducting fundamental changes within a country's energy transition through the large-scale implementation of renewables described in the introduction, technology transfer, and policy diffusion ultimately gained fundamental importance in global climate governance. As Mercure *et al.* point out, the economic instrument of carbon pricing alone, for example, is insufficient to dismantle the institutionalized lock-in effects created by existing fossil technologies and their operators. Instead, "(...) it must be combined with technology subsidies, FiTs, and regulations. This can be ascribed largely to the inertia of diffusion and contrasts with the neoclassical environmental economics view that pricing the externality generates the desired outcome most efficiently" (Mercure *et al.*, 2014, p.695). Combining various technological and political factors and their right application to distinct localities is important for decarbonization. The UNFCC has highlighted the importance of *technology transfer*[10] by creating the technology transfer framework (as a contribution to implementing the conventions' Article 4, Paragraph 5 in 2001) (UNFCCC, no date: a).[11] According to Wilkins,

[10] According to Wilkins "Technology transfer can be defined as the diffusion and adoption of new technical equipment, practices and know-how between actors (e.g., private sector, government sector, finance institutions, NGOs, research bodies, etc.) within a region or from one region to another" (Wilkins, 2002, p.43).

[11] During the 13th session of the Conference of the Parties at Bali, the UNFCCC recognized the fact "(...) that there is a crucial need to accelerate innovation in the development, deployment, adoption, diffusion and transfer of environmentally sound technologies among all Parties, and particularly from developed to

technology transfer and diffusion in the field of RE need to consider various aspects to be applied within a specific local context. Besides being affordable, accessible, and sustainable, specific technologies must also be accepted at the place of their implementation. Transferring technologies thereby refers to an encompassing concept that includes information and knowledge for production, installation, operation, and maintenance to sustainably enable the local usage of technologies (Wilkins, 2002, p.44).

To analyze the transfer processes, *policy transfer* and *policy diffusion* emerged as key theoretical approaches in IR. *Policy transfer* has been defined as:

> "(...) the process by which knowledge about policies, administrative arrangements, institutions, and ideas in one political system (past or present) is used in the development of policies, administrative arrangements, institutions, and ideas in another political system" (Dolowitz & Marsh, 2000, p.5).

Stone states that policy transfer can occur through the transfer of: 1. Policy ideals and goals, 2. Institutional settings and structures, 3. Regulations, administrative and judicial elements, 4. Ideational and ideological perspectives, 5. Personnel, consultants, and experts (Stone, 2012, pp.485–486).

According to Busch and Jörgens, the idea of *policy diffusion,* on the other hand, can be defined:

> "(...) as a process by which information on policy innovation is communicated in the international system and adopted voluntarily and unilaterally by an increasing number of countries over time. Contrary to bilateral or multilateral forms of policy coordination, reciprocity plays no role in diffusion processes. Individual states take over the policies of other states unilaterally,

---

developing countries, for both mitigation and adaptation", and "(...) that the immediate and urgent delivery of technology development, deployment, diffusion and transfer to developing countries requires suitable responses, including a continued emphasis by all Parties, in particular Parties included in Annex I to the Convention, on enhancement of enabling environments, facilitating access to technology information and capacity-building, identification of technology needs and innovative financing that mobilizes the vast resources of the private sector to supplement public finance sources where appropriate" (UNFCCC, 2008, pp.12–13).

> unconditionally and without expecting other states to do the same" (Busch & Jörgens, 2012, pp.222–223).[12]

The process of diffusion, as such, then takes place through different channels. According to Rogers *et al.*, diffusion at large represents a process of communication, which has been studied across various scientific fields (ranging from geography to social sciences and economics and marketing). Research on innovation diffusion processes[13] takes the aspect of time as a distinct variable into account. This specific factor is particularly important concerning technological innovations to fight GHG emissions. This is due to the distinct timespan—from first contact until the possible application of innovation is made (the *innovation-decision period*)—the profoundness of adopting new technologies at an early stage in comparison to other actors (*innovativeness*), and the velocity of an innovation's implementation (*rate of adoption*). A common denominator represents the fact that diffusion of innovation deepens our comprehension of how social changes come about and how technology influences these processes. Out of this insight results in the necessity to study the potential and consequences of technological innovations and their implementation at the local level (Rogers *et al.*, 2009).

As these explanations have shown until now, social changes—which include how we think about energy consumption and how to reduce GHG emissions—are largely induced by innovations. Learning from developments and policymaking elsewhere brought about the understanding that in today's globalizing world economy, the recognition of non-state actors needs to play an

---

12 In international relations, *policy diffusion* according to Stone "describes a trend of successive or sequential adoption of a practice, policy or programme. The 'diffusion' literature suggests that policy change occurs by osmosis; something that is contagious rather than chosen. It connotes spreading or dispersion of models or practices from a common source or point of origin" (Stone, 2012, p.484).

13 According to Rogers, *innovation* can be "(...) defined as an idea, practice, or object that is perceived as new by an individual or another unit of adoption. An innovation presents an individual or an organization with a new alternative or alternatives, and new means of solving problems. However, the probability that the new idea is superior to previous practice is not initially known with certainty by individual problem solvers" (Rogers, 2003, p.xx).

increasingly important role in transferring and diffusing ideas, norms, and technologies. Societal actors and private companies are at the forefront of bringing about societal change. This type of learning should be understood as a collective instead of an individual process. Finally, the special relations between agency and structure and the societal changes induced by learning within these settings must be understood as a comprehensive process (Grin & Loeber, 2007, p.214). Although nation-states continue to be the legitimate actors in negotiating internationally, large parts of policy diffusion and innovation originate outside state-bound and international decision-making. Such a vision gave rise to the thinking in terms of "sovereignty regimes" (Agnew, 2005) beyond the "methodological nationalism" (Stone, 2012, p.490) of the Westphalian state system, in which national governments are not the only or even the main drivers of policy diffusion (Cutler, 2003; Hall & Biersteker, 2004). For example, transferring information, ideas, and technologies within transnational networks can be just or even more effective than between state governments (Stone, 2012, p.491).[14]

Consequently, the emergence of private actors, such as transnational companies, has increasingly gained prominence in policy and knowledge diffusion (Stone, 2001). While the nation-state and state sovereignty have never been as absolute as grand theories of international relations might suggest, the current wave of globalization alongside the privatization in the 1990s and early 2000s brought about new dynamics and were, as Cutler argues, "relocating the boundaries between public and private authority in international commercial relations and creating new opportunities for private, corporate actors to exercise power and influence" (Cutler,

---

14 Arguing in a similar fashion, Gandenberger points out that the resources by nation states are simply too limited to induce significant impacts in other regions of the world. He argues that government funding for policy and technological diffusion alone is insufficient to cause positive spill-over effects on development. Instead, companies of the OECD world, which mainly hold the intellectual property and ownership rights on technological innovations have to be willing to share their knowledge especially with countries of the Global South (Gandenberger, 2015, p.1). This counts especially when referring to the theme of this book and the transfer of ideas on renewable energy technologies and blockchain to contribute to global CO2 emissions reductions.

2003, p.1). According to Susan Strange (1996), these developments have strengthened the decision-making influence on behalf of private entities.[15] In the wake of these developments, private entities such as companies became recognized as "private authorities" and have increasingly started to provide governance services on a national and global scale (Hall & Biersteker, 2004).[16] Stone argues that today, "the role of business in standards setting is well established. In the field of environmental governance, especially Europe, both green and business interest groups have played prominent roles in the advocacy and dissemination of voluntary agreements, ecolabels or certification" (Stone, 2012, p.491). As a consequence, private entities contributed to the creation of "private international regimes" (Cutler, 2004, p.23).

According to Strange, the very technological innovations, which have mainly been neglected in social science and international relations research, contribute profoundly to changes in governance observable on a global scale (Strange, 1996, p.7). However, these developments are neither automatic nor linear. Companies have shown to be reluctant to transfer knowledge and innovations (especially those that demand significant investments in research and development) to countries with limited protection for intellectual property rights in fear of losing their competitive edge (Nanda & Srivastava, 2009; Bennett *et al.*, 2001).[17]

---

15 While coordinating market activities was once primarily a domain of the state, now the markets have started to dominate the agenda setting of governments in a variety of policy fields. As a consequence, the power shift toward markets came along with the erosion of authority of national states (Strange, 1996, p.4).

16 According to Hall and Bierstecker, the conceptualization of private entities as authorities refers to the idea that "(...) private sector markets, market actors, non-governmental institutions, transnational actors, and other institutions can exercise forms of legitimate authority. (...) The state is no longer the sole, or in some instances even the principal source of authority, in either the domestic area or in the international system" (Hall & Biersteker, 2004, p.5).

17 More recent studies also point to the fact that countries like China have rather grown out of their importation of innovation state and started to position themselves as leaders within the renewable energy and low carbon technology sector. As a result, technology transfer can now more frequently be observed on the south-south basis and from southern countries to the industrialized North (Urban, 2018).

3.5. Green Theory

The so-called green theory in international relations emerged based on the beforementioned theoretical models. This approach tries to introduce environmentalism as its theoretical concept to the discipline while considering the political analysis around the various aspects of environmental governance. It tries to define sustainability for current and future generations and considers the "tragedy of the commons" that results from the ecological over-shoot and self-interested human behavior as a global phenomenon (Paterson, 2022; Dyer, 2017).

However, up until today, there is – as has been outlined at the beginning of this theoretical chapter – no unified theory of environmental governance. Green theory started to be recognized as a growing concept after the end of the block confrontation and the international recognition that GHG emissions and climate change pose a real threat to human security. Various concepts have since emerged, which, according to Paterson (2022),[18] can be broken down into four concepts. They range from free market environmentalism[19] and social ecology[20] to institutionalist-based environmentalism[21] and bio-environmentalism-inspired green political thought approaches[22]. As pointed out by Paterson, the main observable divide of environmentally-focused international relations theories can be found between *Environmentalism* as well as *Green Political Thought*. He argues that:

> "(…) environmentalists accept the framework of the existing political, social, economic and normative structures of world politics, and seek to ameliorate environmental problems within those structures, while Greens regard those structures as the main origin of the environmental crisis and therefore contend that they are structures which need to be challenged and transcended. (…) As is obvious from even the most cursory literature survey of the

---

18 While the distinction of the four concepts of "Green Theory" have been outlined by Paterson (2022), the literature references in the footnotes below corresponding to each of these four concepts in this text have been the result of my choosing.

19 See for example: (Eckersley, 1993; Anderson & Leal, 2001).

20 See for example: (Lejano & Skokols, 2013; Bookchin, 2006; 1996; Ungar, 2011).

21 See for example: (Hoffman & Jennings, 2015).

22 See for example: (Dobson, 2007).

> mainstream International Relations literature on environmental problems, the environmentalist position is easily compatible with the liberal institutionalist position outlined most clearly by Keohane (1989a). In fact, most writers within International Relations who write on environmental problems, and who are clearly motivated by the normative concerns adopted by environmentalists, adopt liberal institutionalist positions" (Paterson, 2005, p.236).

So, environmentalist perceive the global ecological challenges to be solvable within the existing structures of the international system and successfully to be negotiated within international institutions such as UNFCC. They therefore hold an anthropocentric, human-centered world view.

In distinction to these perceptions, ecocentric positions shared by green politics are drawn by a holistic worldview, denying human beings' supremacy over other living creatures. According to these approaches, economic and industrial growth (alongside the technological advancements they have generated) has reached the limits of sustainability. Therefore, the current capitalist system, the polluting form of industrialization, and today's consumption-based Western lifestyles are incompatible with sustainable development (Tayyar & Gökpınar, 2019, pp.167–168). Being, therefore, part of the critical approaches to IR theory, Paterson argues that the Marxist conception of structural inequality and dependency by certain regions in the global capitalist system is dominant in green political thought. Focusing furthermore on the distinction of values proposed by critical theories, feminism, and post-structuralism, the potentially despotic concentration of power and the main driving forces behind the ongoing global homogenization are being rejected.

In contrast, an argument favoring distinct and unequal global developments leading to different social and economic realities is upheld. Resulting from these perspectives is a notion of a global community characterized by diversity and thus focused on the particularities of local settings. Linked to such a vision is subsequently a general questioning of the nation-state as the sole or even most efficient actor to organize, represent, and communicate the interests of the global population in the international arena and negotiate all-encompassing solutions. So, the state-based international system and the authority structure resulting from its political-economic configuration are the core reasons for inefficient and partially unjust global climate change policies. Consequently, green political thought entails a normative, explanatory, and emancipatory foundation in that it is goal-oriented and considers the rethinking of these status quo components essential for a successful transformation of global environmental governance. Bio-environmentalism's decentralist ideas have gained prominence through the "think global, act local" frame, which named community-based political and self-reliant economic actions to solve global environmental affairs (Paterson, 2022).

Furthermore, Eckersley (2012) calls for changes in the representation of international actors in global climate dialogs. To best negotiate effective global climate strategies, neither an "inclusive multilateralism" nor an elite-based "exclusive minilateralism" are ways to sort out the key actors for successful and legitimate climate talks. Instead, an "inclusive minilateralism" should be considered, which is based upon a '"(…) common but differentiated representation,' that is, representation by the most capable, the most responsible and the most vulnerable" (Eckersley, 2012, p.26). So, different institutional levels bring distinct political and economic bargaining power to the negotiation table. At the same time, a reciprocal dependency between these actors exists due to their specific knowledge. Incorporating such knowledge into climate policies is essential to unfold an effective climate government's potential. According to Eckersley's understanding, the global Covid-19 pandemic highlighted the vulnerable structure of present-day global environmental governance. Therefore, the "critical conjuncture" of the post-pandemic economic reactivating phase should stimulate green investments while abandoning the most emission-intensive industries step by step. Keeping the economic effects of the pandemic on certain industry sectors in mind, economic stimulus packages should be oriented toward creating local and regional development strategies to generate growth, employment, and consumption based on sustainable strategies, especially in areas related to a state's critical infrastructure. By doing so, communities would be better prepared for possible future pandemics and trigger a higher level of community engagement, generate cooperative forms of precaution, and allow for reflexive learning. Such thinking entails the possibility of focusing less on growth determined on a monetary assessment basis but rather opens options for new forms of state-society complexes with positive effects in ecological, justice, and human security terms (Eckersley, 2021).

With regard to the implementation of adequate RE strategies, therefore, private, public, and non-governmental actors need to be brought together to work out institutional best practice arrangements. Although the Westphalian state system remains essential in this context to allocate financial resources to pass adequate welfare programs and economic promotion laws, nation-states are badly equipped to lead a sustainability-based transformation without recognizing distinct local political-economic configurations and social dynamics. Consequently, the idea of including analytical levels beyond the state to steer globally sustainable policies from a bottom-up perspective that considers distinct local settings is of importance.

It remains, therefore, to be seen how far the local implementation of blockchain technology represents a potential strategy to curb GHG emissions and is being accepted as a technological but also a policy tool to enhance the greening of the energy sector by actors from different social backgrounds.

3.6. Conclusion

This chapter began by asking why there is no unified theory of global environmental governance. It started by showing that an increasingly important global governance approach exists, which offers theoretical tools to analyze environmental challenges brought about by key issues such as rising global GHG emissions.

Regime theory, as a so-called 'problem-solving theory,' shows that international cooperation and the creation of international regimes such as the one on climate change represents an opportunity to coordinate policies between states and to benefit from collective action. However, this theory highlights the challenges of policy coordination resulting from diverging interests between states and fears of relative political and economic losses. In contrast to classic security-related issues, the threat of climate change appears yet to be too abstract by some states in order for them to agree on sovereignty transfer and the creation of efficient international climate regimes. Effective political concessions to maintain global warming well below the intended two-degree Celsius level proposed by the Paris Agreement are only visible to a limited extent.

Policy learning and diffusion approaches have thus emerged as theoretical concepts to analyze the transfer, implementation, and materialization of political ideas, conceptions, and technological developments emerging in certain geographical areas in one part and applied to distinct local, regional, or national sociopolitical configurations in other parts of the world. Given the competitive logic of the global economy and existing challenges to guarantee intellectual property rights in many countries, however, private actors such as companies remain very reluctant to share their knowledge fully.

Finally, green theory, a theoretical concept with liberalist and critical strands, differs in its perspectives on overcoming global challenges such as climate change in a world of competing nation-states. Discussions primarily focus on the differences between the anthropocentric (human-centered) approaches, which advocate the finding of solutions to the climate crisis through negotiations within international institutions, and the ecocentric (holistic) theories of green political thought that propose the inclusion of actors from the different social and political level into climate politics to solve global environmental problems.

In the following, the CO_2—climate change and energy change—nexus shall be explored, the prime reasons contributing to the emergence of emissions outlined, and possible approaches to reduce these GHG emissions in the energy sector effectively will be presented.

4. Climate Change

4.1. CO_2—Climate Change Nexus.

In the words of the UN Secretary-General António Guterres:

> "Climate change is moving faster than we are—and its speed has provoked a sonic boom SOS across our world. If we do not change course by 2020, we risk missing the point where we can avoid runaway climate change, with disastrous consequences for people and all the natural systems that sustain us. (…) The high level of carbon dioxide in the atmosphere is making rice crops less nutritious, threatening well-being and food security for billions of people" (United Nations Secretary General, 2018).

According to the 2023 Synthesis Report of the IPCC Sixth Assessment Report, the global surface temperature was 1.09 degrees Celsius higher between 2011 and 2020 than between 1850 and 1900. Human-caused surface temperature increases from 1850–1900 to between 2010 and 2019 lie between 0.8 and 1.3 degrees Celsius (IPCC, 2023, p.4).

Within this context, it is important to highlight the fact that the climate is changing because of a variety of different factors. Elements to be considered are alterations in the earth's orbit, changes in the sun's intensity, and currents of the world's oceans (Florides & Christodoulides, 2008, p.391).

According to the IPCC, anthropogenic greenhouse gas (GHG) emissions contribute profoundly to the large increases in atmospheric gas concentrations. The three most prominent gases are carbon dioxide (CO_2), methane (CH_4), and nitrous oxide (N_4O) (IPCC, 2014, p.4). The IPCC states that the historical cumulative net CO_2 emissions between 1850 and 2019 range at 2040 ± 240 $GtCO_2$. Fifty-eight percent were emitted between 1850 and 1989, while 42 percent occurred between 1990 and 2019. The atmospheric CO_2 concentrations in 2019 were 410 parts per million and have been the highest in the last two million years. Methane concentration levels were 1866 parts per billion, and nitrous oxide at 332 parts per billion. Both levels were higher than at any time in the last 800,000 years. Average greenhouse gas emissions between 2000 and 2019 displayed the highest measured level, although the global growth rate was 1.3 percent compared to 2.1 percent between 2000 and 2009. Also, interesting is the fact that reductions in CO_2 fossil fuel and industry emissions (CO_2-FFI) resulting from the lowering of the GDP's energy intensity and the carbon intensity of energy have shown to be lower than emissions stemming from industrial, energy, supply, transport, agriculture and construction activities (IPCC, 2023, p.4). So, economic and population growth contribute most to the increase in CO_2 emissions (IPCC, 2014, pp.4–5).

If continued at the current pace, global warming will likely increase by 1.5 degrees Celsius between 2030 and 2053. According to the IPCC, anthropogenic global warming is rising at around 0.2 degrees Celsius every decade due to past and present emissions. Warming caused by anthropogenic emissions from the pre-industrial periods up until today shall persist for centuries while contributing to the climate system's alteration (IPCC, 2018, pp.4—5). Focusing on the long-term damages, Solomon points out that:

> "(...) irreversible climate changes due to carbon dioxide emissions have already taken place, and future carbon dioxide emissions would imply further irreversible effects on the planet, with attendant long legacies for choices made by contemporary society. Discount rates used in some estimates of economic trade-offs assume that more efficient climate mitigation can occur in a future richer world, but neglect the irreversibility shown here" (Solomon *et al.*, 2009, p.1709).

Although voices suggest that a clear correlation between CO_2 emissions and global warming cannot be properly proven (Florides & Christodoulides, 2008), a general scientific consensus exists around the fact that carbon dioxide has already irreversibly affected the global climate. The IPCC also calls for immediate mitigation measures to maintain the atmospheric temperature increase at the intended 1.5 degrees Celsius compared to the pre-industrial level (IPCC, 2023).

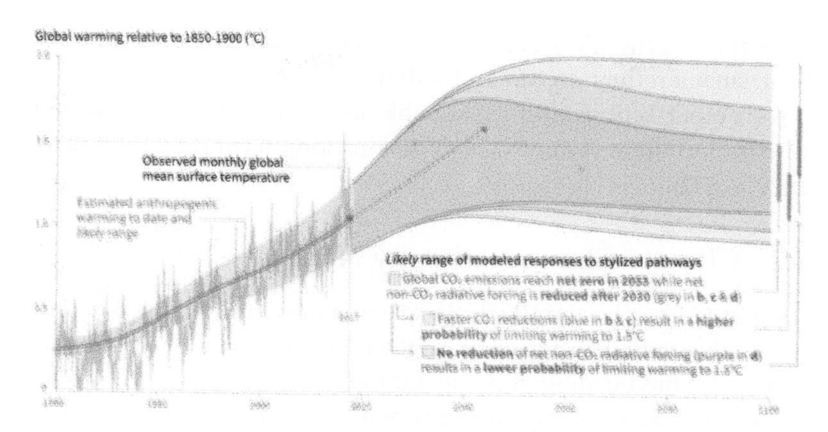

Figure 1: IPCC (2018, p.6). "Cumulative emissions of CO_2 and future non-CO_2 radiative forcing determine the probability of limiting warming to 1.5°C."

As the graph demonstrates, faster and more significant CO_2 reductions (as reflected in the b & c lines) from 2017 onwards shall result in a higher possible impact regarding the global temperature's stabilization. The longevity of CO_2 in the atmosphere and the even stronger need to act urgently were best displayed during the Covid-19 pandemic. Although large parts of the world were in a complete lockdown, global CO_2 emissions hit a new record high in May 2020. As atmospheric CO_2 leveled above 417 parts per million, the global warming effect remained strong despite a significant decrease in CO_2 production during the crisis (Financial Times, 2020).

The global pandemic, however, triggered a significant reduction in energy consumption and CO_2 reductions stemming from fossil fuel consumption. In 2020, global CO_2 emissions dropped by 2.3 billion tonnes, which represents a decrease of 6.4 percent (Tollefson, 2021). The first half of 2020 showed a reduction of 8.8 percent in CO_2 emissions compared to the first six months of 2019. However, from July of that year onwards, the lockdown effects on CO_2 emission reductions decreased as some countries started to lift some of their restrictions on social distancing measures (Liu *et al.*, 2020).

Without fundamental structural changes within the global economic system, the current transportation methods, and most importantly, the global energy production and consumption pattern, such a reduction can and will not be achieved. Post-Covid-19 government actions and sustainable incentive schemes are therefore important to reduce CO_2 emissions long-term (Le Quéré *et al.*, 2020, p.652).

4.2. Climate Change – Energy Nexus

As pointed out by the IPCC, decarbonization can only be successfully achieved when focusing on the climate change-energy nexus. Primarily, the burning and consuming of fossil fuels have proven to be the major source of anthropogenic GHG emissions (IPCC, 2012, p.7). This observation is even more challenging when taking into account that the primary energy demand continued to grow by 5.8 percent in 2021, thereby overrunning the 2019 levels by 1.3 percent. Oil consumption rose by 5.3 million barrels/day, yet remained 3.7 million barrels/day beneath 2019. Carbon dioxide emissions from energy usage, industrial processes, flaring, and methane (measured in CO_2 equivalents) grew by 5.7 percent, or 39.0 $GtCO_2$, in 2021. Energy-related CO_2 emissions rose by 5.9 percent to 33.9 $GtCO_2$ and are, therefore, close to the 2019 level.

Additionally, the demand for natural gas increased by 5.3 percent, breaking the 4 Tmc level for the first time and thus displaying the continuing shift from petroleum toward producing and consuming natural gas as the prime global energy source. Gas supply, however, reached the lowest growth level since 2015 with a volume of 516 Bcm. Even more disturbing in terms of CO_2 reduction efforts remains the fact that coal production (440 Mt.) and consumption (160EJ) increased in 2021. The latter went up by more than 6 percent, thus reaching a value above the 2019 level and showing the highest rise since 2014. China and India were the biggest producers, with a share of around 70 percent, and mainly consumed this resource domestically. However, despite its efforts to become carbon neutral in the coming decades, Europe's coal consumption increased for the first time in almost ten years. Finally, global electricity generation increased by 6.2 percent, contributing 36 percent from coal, 22.9 percent from natural gas, and 10.2 percent from RE sources (mainly wind and solar). For the first time, the latter is higher than the contribution stemming from nuclear energy (BP, 2022, p.3).[23] As these statistics highlight until now, there exists a strong incoherence between the continuous consumption of fossil fuels and the intended goals to reach a significant reduction of CO_2 emissions.

Drawing on Yergin's (2006) definition of energy security, the argument is made that a growth-based global economic logic demands constant access to affordable, reliable, and unconstrained energy sources. However, the additional aspect of sustainability—necessarily based on promoting RE sources (Elkind, 2010; Casola & Freier, 2018)—has and continues to be neglected by large parts of the world's policymakers. Beyond the occasional reference to renewable resources as a viable tool to reduce GHG emissions, energy security, and climate change-related issues are treated as two distinct policy fields when observing the legislation on the national

---

23 Coal-fired power plants showed the highest contributions to emissions in 2018. Electricity generated from these plants contributed 30% to global CO_2 emissions (IEA, 2019). Even EU countries, which had previously positioned themselves as leaders in the fight against GHG emissions, prolonged the generation of electricity from coal. Initially, the EU commission recommended the closing date for coal fired power generation for 2038, if the energy economic conditions to do so shall exist (BMWi, 2019a, p.64). However, neither the latter date, nor the vaguely announced measures for its reduction until the year 2030 are sufficient for the energy sector to make an adequate contribution to climate protection (BMWi, 2019a, p.119). Meanwhile, the EU Climate Law has been passed and obliges the EU member states to achieve a reduction in GHG emissions of 55% in 2030 relative to the level in 1990. Furthermore, the EU target is to reach climate neutrality by 2050 (European Commission, 2023e). To achieve this objective, emissions reductions in the electricity sector of 700 Mt, or of 71% relative to the year 2015 are necessary. Attaining this goal is expected to be achieved by the EU Commission, if coal generation is reduced to less than two percent until the year 2030. According to the German think tank Agora Energiewende and energy consultancy Enervis, a policy mix consisting of EU- and national policies is necessary. To phase out the additional coal use of the "Coal-6" countries in the EU, a 100 GW expansion of renewable energy (mainly wind and solar) and an additional implementation of 15GW of gas power plant capacity is expected to be needed during the renewables' implementation phase (Agora & Enervis, 2021, p.3). Although, Russia's invasion of Ukraine in February of 2022 led to a severe European energy crisis, strongly impacted the EU's energy policies and led coal usage to be considered as a back-up tool by a number of countries, the REPowerEU plan to accelerate renewable energy implementation, energy savings, and the ETS carbon pricing undermine a general return to the use of coal (Herzog, 2022).

and intergovernmental levels in different parts of the world (Casola & Freier, 2018). In the same line of contradiction to achieve a reduction of CO_2 emissions, the politics to abandon nuclear power plants in Germany and a few other European countries without having the technological capabilities available to replace these low-carbon energy generators have been subject to criticism (Sinn, 2017). According to this argument, political decision-making tends to orient itself on short-term public opinion changes after catastrophic events such as the nuclear reactor explosion at Fukushima. While the resulting public fears are very much understandable, the socio-economically complex realities of energy governance are insufficiently taken into account. Second, a conflict of objectives between reducing CO_2 emissions and environmental considerations exists. So, while nuclear power as the second lowest CO_2 emitting energy source in the European Union has been abandoned completely in Germany in 2023 and will be in other European countries, the environmental and social consequences are expected to be high due to the currently lacking visionary alternative low-carbon resources (Jarvis *et al.*, 2022).[24] On the global level, however, nuclear power continued to grow by 2.4 percent and, in 2018, displayed its highest growth rate in eight years. China and Japan account for the biggest increase (BP, 2019, p.2).

The abovementioned statistics highlight that four distinct decarbonization indicators linked to the final energy consumption must be considered to create an energy system transformation with significant reductions in GHG emissions. According to the IPCC, these are: "(…) limits on the increase of final energy demand, reductions in the carbon intensity of electricity, increases in the share of final energy provided by electricity, and reductions in the carbon

---

[24] In Germany, the last three nuclear power reactors had a total net capacity of 4055 MW(e) and were phased out on April 15, 2023 (IAEA, 2023). On the EU-27 level, electricity produced by nuclear power plants accounted for around 25.2% in 2021, with then 13 countries still having active power plants running. However, electricity generated from nuclear power plants dropped by 20% between 2006 and 2021 within the EU-27 (Eurostat, 2022). The overall social cost of the nuclear phase out is estimated to be around $3 to 8 $bln. in health-related cost, annually (Jarvis *et al.*, 2022).

intensity of final energy other than electricity" (IPCC, 2018, pp.129–130).

While these goals make changes in consumer behavior and a subsequent awareness among the members of the world's societies necessary, the most important aspect remains a shift in perspective toward the development and use of RE resources. As we have seen until now, low emissions from traditional energy sources only stem from the nuclear sector and leave renewables as the only remaining alternative. The International Renewable Energy Agency (IRENA) has defined RES to include "all forms of energy produced from renewable sources in a sustainable manner, including bioenergy, geothermal energy, hydropower, ocean energy, solar energy and wind energy" (IRENA 2013, cited in Sustainable Energy for All, 2014, p.194). According to IRENA, at the end of 2022, the globally installed power capacity from RE resources amounted to 40 percent. With around 295 GW, 2022 saw the largest-ever increase in RE capacity. One hundred and ninety-two GW of solar and 75 GW of wind energy were implemented. The latter, however, was less than the 111 GW capacity added in 2020. To meet the climate objectives and limit temperature increases to below two degrees Celsius, the worldwide renewable expansion needs to be above 1000 GW annually until 2050 (IRENA, 2023).

Within this outlined context, it must be mentioned that (according to a study based on OLS and GMM estimation techniques conducted by Adams and Nsiah) both non-renewable and RE resources contribute to CO_2 emissions in the long run. This environmental impact of renewables is mainly associated with the sporadic character and subsequent fluctuations of RE generation and consumption due to a lack of storage mechanisms (Adams & Nsiah, 2019). Furthermore, it has to be pointed out that while renewables are merely free of CO_2 emissions when operated, one has to consider emissions that result from the production of plant installations. According to a study conducted for the German parliament (Federal Parliament, 2007), not only emissions stemming directly from energy operations should be considered. Hydropower plant constructions, for example, lead to GHG emissions by producing massive concrete and creating artificial damned lakes. This contributes to the rotting of natural plants, which then might release methane (CH_4). The latter is a significantly more potent greenhouse gas than carbon dioxide (Le Fevre, 2017, p.9).

However, wind use, especially solar power, is more important for this book's analysis. Wind energy-related CO_2 emissions occur within the value chain during wind towers' installation, maintenance, and decommissioning process. Onshore wind's life cycle carbon emissions range between 3 to 45g CO_2eq/kWh. If built on forested peatland, these emissions can increase from 62 and 106g CO_2 eq/kWh. On the other hand, carbon emissions for offshore wind usage range between 7 to 23 g CO_2eq/kWh (Thomson & Harrison, 2015, pp.2, 16).[25]

Solar energy, where solar cells are used to transform solar radiation into electricity, causes emissions in its production cycle. Depending on the distinct types of solar cells, so-called constructed silicon must be thoroughly cleaned before melting and hardening into crystals and cutting into thin wafers. These processes come along with intense energy insertions. Thin-layer solar cells and silicon cells with lesser clarity are possible alternatives. Comparable to the case of wind energy, sun radiation and the exact horizontal or vertical sun orientation at distinct locations play an important role in the electrification process (Federal Parliament, 2007, pp.17–19).

Generally, however, the main challenges to reducing CO_2 emissions are related to the consumer consumption pattern and the distinct energy matrix at a given location. So, the carbon intensity of the grid load depends on the fluctuations within the electricity production cycle. An hourly measure of the carbon metrics is suggested to determine these fluctuations and the interaction of renewables with the grid mix. By applying these methods, accurate

---

[25] Furthermore, the electricity generation output of wind energy is location-dependent. Taking the law of Betz into account, the level of energy generation is increasing according to the cubic power of wind speed. So, in wind active areas, the doubling of the velocity of wind can amount to an eightfold level of electricity generation. A tripling of wind velocity can lead up to a 27 times higher level of energy generation (Federal Parliament, 2007, pp.16–17). Wind speed therefore represents the independent variable, whose intensity manifests itself within a defined level of CO_2 emissions.

control signals will allow for the most suitable feed-in of additional renewable resources. For example, this can be achieved through a combination of solar and wind energy (Chalendar & Benson, 2019). Qi *et al.* argue in a similar vein when highlighting that supply-side policies targeting RE increase alone do not show significant effects on reducing CO_2 emissions. This is especially true when RE implementation does not come along with reducing fossil fuel usage in other industrial sectors. In this case, the authors describe an "offsetting leakage effect." Most effective regarding the overall increase of renewables would be a cap-and-trade system and, therefore, the pricing of externalities caused by fossil fuels on the environment (Qi *et al.*, 2014, p.68).[26] The main challenge regarding the EU ETS has been attributed to the low prices as a result of the high availability of certificates until its sudden increase in 2022. In this argument, Sinn highlights that national attempts to reduce CO_2 emissions are prone to fail. One of the key elements of the "green paradox" is attributed to the fact that ETS certificates liberated by countries like Germany, who have made strong efforts to increase its share of energy generation from renewable resources, contributed to a higher availability and price reductions of these certificates on the European level and thus decreasing concerted actions to minimize emissions by other European countries (Sinn, 2012).

As a result, the initial prices of certificates listed between €20 and €30 per ton of CO_2 in 2008 dropped to €3 to €4 by July 2013 (Kirsten, 2014, p.309). In May 2020, the price per certificate was traded at €26.95/t CO_2 at the Leipzig-based European Energy Exchange (Business Insider, 2020) and rose above €100/t CO_2 for the

[26] The success of the market-based mechanisms to cap emissions is subject to academic and public debate. While emissions savings due to the EU ETS for example is being agreed upon (at a range between 40–80 $MtCO_2$ in average per year) and has shown to be more effective than any other environmental policy instrument, it has not yet inspired further large-scale investments into low-carbon technologies. The EU ETS has instead shown to be merely a tool to generate awareness among large consumers such as major companies and impacted their decision-making against investments into CO_2-intensive technologies (Laing *et al.*, 2013, p.25).

first time in February 2023 (Trading Economics, 2023).[27] Against initially low trading prices for certificates, Bayer and Aklin argue that the EU ETS, which regulates around 45 percent of all carbon emissions stemming from energy production and industrial pollution, has indeed led to CO_2 emission savings of around 1.2 bn tons of carbon dioxide between the 2008 and 2016. According to this study, the reductions equaled a 3.8 percent drop in all EU-wide emissions relative to the non-existence of such mechanisms despite comparatively low-carbon emission prices (Bayer & Aklin, 2020, p.8804). So, although a concerted strategy would be more effective, national efforts alone have also been shown to cause certain effects. However, international carbon trading has to be aligned with national emission reduction targets and other national policies. As a result, for additional national CO_2 reduction policies to unfold their true potential, governmental decision-makers must ensure that these policies' design complements and does not diminish the effectiveness of international CO_2 frameworks (Kirsten, 2014, p.309).

4.3. Conclusion: Energy – CO_2 Nexus

As shown until now, the fight against global climate change cannot be effectively pursued if it does not come with a policy shift toward aligning national environmental policies, enabling a gradual but steady shift toward transforming the world's energy matrix. Most countries that have signed the 2015 Paris Accord named the expansion of their RE capacities a key element within their Nationally Determined Contributions (NDCs) to comply with GHG reduction objectives. According to IRENA, RE expansion combined with EE strategies must be the prime focus to reduce CO_2 emissions from

---

27 According to Gerlagh et al., the price spike of ETS certificates from €5 t/CO_2 to above €90 in 2021 is linked to the cancellation mechanism introduced to the EU emissions trading scheme alongside the market stability reserve (MSR) in 2018. The latter led to high cancellations of emissions allowances. According to the authors, this indicates a necessary steepness of the emissions pathway. The flattening of this very pathway induces substantial reductions in cumulative emissions and higher prices for certificates. Consequently, the current debate on the inefficiency of the ETS due to low certificate prices has shifted toward a discussion on the possible negative impacts of high ETS prices (Gerlagh *et al.*, 2022).

burning fossil fuels drastically. To do so, the agency calls for a broad and human-centered approach to comply with the Paris Treaty's climate objectives and, therefore, an encompassing design to lower CO_2 and other GHG emissions. IRENA regards the human-security approach as integral to sustainable development (IRENA, 2019). Such a position is in line and can be explained by global governance, regime theory, and green theory approaches, as outlined in the chapter on international relations theory. However, a consensus to create effective contributions among state, private, and societal actors has to be found. The continuously growing GHG emissions serve as a strong indicator that such consensus does not yet exist.

So, to find a suitable solution for the energy sector as the prime contributor to global GHG emissions, appropriate adjustment measures must be implemented at the community, regional, federal, and international levels. Consequently, IRENA links the challenge of reducing CO_2 emissions and balancing demand and supply to the digitalization of the RE sector. Innovations in smart technologies such as blockchain, artificial intelligence, DERs, information-, and communication technology, smart contracts, and the Internet of Things (IoT) are at the heart of this concept and demand a further decentralization of energy generation (IRENA, 2019, p.28).

Before going deeper into the analysis of these technological innovations, it shall first explore how far technological innovations such as blockchain can be derived from the international climate contracts as a possible bottom-up strategy for the energy sector to fight global GHG emissions.

5. Normative International Framework

5.1. International Climate Contract and Climate Technologies

In the late 1970s, the global community began to express increasing concern about climate-related issues, leading to a significant milestone in the form of the first World Climate Conference in 1979. This event marked the initiation of a worldwide movement advocating for environmental protection, culminating in adopting the UNFCCC[28] in 1992, which subsequently came into force in 1994.

Looking ahead, it can be observed that 198 nations have come together under the convention, united by the shared goal of averting harmful human interference in the climate system (UN Treaty Collection, 2023). This substantial participation demonstrates the widespread acknowledgment and dedication toward confronting the challenges of climate change and acknowledging its potential ramifications.

The UNFCCC represented a significant breakthrough, recognizing climate change as a worldwide concern and highlighting the necessity of implementing coordinated strategies to alleviate its adverse effects. However, the convention maintained a broad scope without delineating specific actions (Urrutia Silva, 2010, pp. 603–605). As a result, it has primarily been complemented through the Kyoto Protocol, the Doha Amendment, and the Paris Agreement.

In 1997, participating nations endorsed the Kyoto Protocol to concentrate their endeavors on curtailing greenhouse gas emissions through well-defined target goals. This protocol, which established a more precise framework for tackling the complexities presented by climate change, came into force in 2005 and currently has 192 parties.

In contrast to the convention, the Kyoto Protocol introduced binding obligations for states, requiring them to adhere to specified

---

[28] The UNFCCC was approved during The United Nations Conference on Environment and Development (UNCED), also known as the '*Earth Summit*', held in *Rio de Janeiro* (Brazil) in 1992.

targets for reducing emissions. Specifically, 38 parties listed in Annex I of the Convention[29] (37 states and the European Union) are included in Annex B of the Kyoto Protocol, which outlined precise reduction objectives for the initial commitment period (2008–2012).[30] As emphasized, these targets aimed at achieving an average five percent reduction in emissions compared to 1990 levels over the course of those five years (UNFCCC, no date: b).

To achieve these objectives, nations can implement their own domestic strategies. In addition to local measures, the Kyoto Protocol introduced three market-based mechanisms that established the so-called carbon market. These mechanisms aim to promote technology transfer for emission reduction and can be classified into three categories: the clean development mechanism (CDM), joint implementation (JI), and emission trading (ET).

The clean development mechanism, as outlined in Article 12 of the Kyoto Protocol, aims to support non-Annex I Parties in their pursuit of sustainable development while limiting their greenhouse gas emissions. To this end, Annex I countries can undertake emission reduction projects in developing nations and receive certified emission reduction units in return. These units can be utilized to fulfill their own quantified commitments. Notably, the CDM requires projects to undergo a thorough approval process by designated national authorities to ensure that the emissions reductions achieved are additional to what would have naturally occurred.

Regarding the JI mechanism, it enables countries listed in Annex I of the Convention (Annex B of the Protocol) to participate in emission reduction or removal projects implemented in another Annex I country. By doing so, they earn emission reduction units

---

[29] Currently, there are 43 parties included in Annex I, which consists of industrialized nations that were part of the Organization for Economic Co-operation and Development (OECD) in 1992. Additionally, countries undergoing economic transitions (the Russian Federation, the Baltic States and some Central and Eastern European States) and the European Union are also encompassed within this group (UNFCCC, no date: h).

[30] Those objectives address the emissions of the six main greenhouse gases, namely, carbon dioxide (CO_2), methane (CH_4), nitrous oxide (N_2O); hydrofluorocarbons (HFCs); perfluorocarbons (PFCs); and sulfur hexafluoride (SF6). See the information: (UNFCCC, no date: c).

that can be used to fulfill their own targets set forth by the Kyoto Protocol. This mechanism is often explained by saying that "joint implementation offers Parties a flexible and cost-efficient means of fulfilling a part of their Kyoto commitments, while the host Party benefits from foreign investment and technology transfer" (UNFCCC, no date: d).

Finally, the ET mechanism allocates a specific number of allowable emissions for each state party. Consequently, the protocol allows countries to sell their unused emissions within their permitted limits to other countries that have exceeded their emission targets. This system operates as a global mechanism for emission compensation. Consequently, a new tradable asset in emission reductions or removals has emerged. As carbon dioxide, the primary greenhouse gas can be traded similarly to other commodities, this phenomenon is known as the carbon market (UNFCCC, no date: e).

Within the carbon market framework, units traded under the protocol, including emission units, removal units, emission reduction units, and certified emission reduction units, must be meticulously tracked and recorded within an official registry system. The international transaction log (ITL) plays a vital role in securely facilitating the transfer of these units between states.[31]

In light of the preceding context, the Doha Amendment to the Kyoto Protocol was adopted by state parties during the 8th COP, serving as the meeting of the parties to the Kyoto Protocol (CMP) held in Qatar in 2012. This amendment established a second commitment period (2012–2020) to advance and intensify national efforts to mitigate greenhouse gas emissions. The amendment came into effect in December 2020 and incorporated new obligations for Annex I parties to the Kyoto Protocol and a revised inventory of

---

[31] Registries send transaction proposals to the ITL, which assesses each proposal against a set of defined checks and responds to registries' transaction proposals by clearing them for further processing. If a transaction proposal is approved, the registry completes the transaction but if the transaction proposal is rejected, the ITL sends an error code indicating which ITL check has failed and the registry terminates the transaction. The full specifications of the ITL system are defined in the Data Exchange Standards (DES), which describes the technical requirements for the communication between the ITL and registries. It also provides the list of all the checks performed by the ITL (UNFCCC, no date: f).

greenhouse gases to be reported. During this second commitment period, parties committed to achieving a minimum reduction of 18 percent below 1990 emission levels within eight years.

Another important turning point in the progression of the international legal framework addressing climate change occurred in 2015 with the signing of the Paris Agreement. This Agreement holds great significance as it establishes four key points.

First, the parties pledge to implement strategies aimed at restricting the increase in global average temperature to below 2 degrees Celsius above pre-industrial levels. Furthermore, efforts are to be made to pursue even more ambitious goals by limiting the temperature increase to 1.5 degrees Celsius.

Second, the Agreement highlights the crucial need to mitigate climate change and adapt to its effects, recognizing that certain consequences are irreversible. As a result, states are urged to undertake adaptive measures to mitigate the adverse outcomes.

Third, the Agreement envisions the establishment of financial mechanisms to support the efforts of developing countries and the most vulnerable nations to reduce greenhouse gas emissions and adapt to the consequences of global warming.

Finally, it must be highlighted that the Paris Agreement also stipulates the establishment of a new framework for technology transfer and capacity building. This framework aims to support developing or underdeveloped countries in pursuing national strategies to achieve their climate goals.

As it becomes evident, technology plays an important role in the global strategy to address climate change-related issues. On the one hand, it is closely intertwined with the mechanisms employed to reduce or eliminate greenhouse gas (GHG) emissions. On the other hand, it is a key component of the capacity-building process to assist developing countries in fulfilling their environmental commitments. Climate technologies, encompassing a wide range of solutions, are utilized to combat the root causes of climate change. These include renewable energies, drought-resistant crops, early warning systems, and sea walls. Additionally, 'soft' climate technologies such as energy-efficient practices and training for equipment usage contribute to mitigating climate change (UNFCCC, no date: e).

The significance of technology in the fight against climate change is so crucial that the UNFCCC mandates state parties to collaborate in developing and transferring climate technologies. To strengthen this process, in 2010, the COP established a technology mechanism that comprises two bodies: the Technology Executive Committee, consisting of 20 experts who offer policy recommendations, and the Climate Technology Centre and Network, responsible for implementing policies and providing technical assistance to member states upon request (UNFCCC, no date: g). Furthermore, Article 10 of the Paris Agreement introduced a technology framework to support the actions facilitated by the technology mechanism.

Capacity-building, involving acquiring knowledge, skills, and resources necessary to tackle climate change challenges, is an essential complement to technology development. It plays a crucial role in facilitating the development and transfer of climate technologies. Therefore, acknowledging its significance, COP 7 held in Marrakech in 2001 approved two distinct frameworks for capacity-building. One framework is specifically designed to support developing countries, while the other is dedicated to assisting economies in transition, and although they share similarities in structure and substance, they differ in specific details.[32]

5.2. Conclusion

This chapter demonstrates that the UNFCCC significantly emphasizes technology transfer and disseminating technical knowledge among nations. The recognition of low-carbon technologies as crucial contributors to GHG reduction is well established.

The Kyoto Protocol primarily focused on achieving explicit and obligatory emission reduction targets. To accomplish this, the concept of technology transfer was derived from three key

---

[32] For more detailed information about the capacity-building frameworks see: (UNFCCC, 2002).

mechanisms: the CDM, JI, and ET. Facilitating technological knowledge transfer to developing countries was deemed essential to enable them to meet the prescribed emission targets.

On the other hand, the Paris Climate Agreement aligns closely with the regime's complex elements described by Keohane and Victor (2011). To address the challenges of decarbonization and environmental preservation, greater flexibility was provided to countries in defining their national targets for GHG emissions reduction.

However, the potential link between blockchain technology and environmental protection remains an open question. Although it has been suggested that the fundamental pillars of future energy systems encompass decarbonization, decentralization, digitalization, and empowering consumers (Andoni *et al.*, 2019, p.144), the subsequent chapters will address the inquiry of whether blockchain platforms can effectively contribute to decarbonization efforts through the distribution of RE.

6. The Digitalization of Renewable Energy

6.1. Contextualization

As shown in the previous chapter, international climate contracts support the transfer of technologies between countries and create an incentive structure to invest in technical low-carbon innovations. As outlined, the energy sector and its necessary decarbonization must be at the center of such a process.

Consequently, IRENA links the reduction of CO_2 emissions in the energy sector to the deployment of smart solutions to enhance the flexibility of RE availability (IRENA, 2019, p.28). The main challenge in both highly industrialized countries, such as in Europe and/or nations characterized by bigger land masses, is to maintain and improve the stability and functionality of the grid. To do so, digital innovations in the field of information and communications technology (ICT) have enabled the development of smart grids (Panahi & Panahi, 2020, p.2). The latter is necessary as the distributed energy generation and the increase of RE sources are impacting traditional energy transmission systems. Being coupled through static conversion systems, reducing production from traditional generators will result in declining units able to provide services for the transmission system. This is said to be a main contributing cause of declining inertia. So, while favored from an environmental standpoint, from a technical distribution system perspective, the large-scale feed-in of photovoltaic and wind energy leads to major challenges. Uncertainties regarding the grid system's performance and the increasing risks related to contingencies can potentially decrease the energy distribution system's network capacity. Additionally, the creation of bi-directional power flows (substituting the traditional unidirectional stream) and the subsequent options to reverse the flow of electricity demand intelligent solutions to replace the traditional pattern of passive distribution management (Ghiani *et al.*, 2018, p.1).

6.2. Prosumer

According to estimations by IRENA:

> "energy-related CO_2 emissions from all sectors totaled 36 Gt in 2015. These need to fall to 13 Gt in 2050 to achieve the REmap scenario, a reduction of 70% compared to the Reference Case, under which emissions are estimated to reach 45 Gt in 2050. Renewable energy could provide 44% of these reductions (20 Gt per year in 2050) (…)" (IRENA, 2017:3).

To meet these GHG reduction targets and fully unleash the potential of the RE markets' growing digitalization, the idea of *prosumption* gained prominence over the last few years (Lowitzsch *et al.*, 2020; Lowitzsch, 2019). The main idea behind this process is that so-called *prosumers* take on a double role as "producers" and "consumers" of energy. According to Parag and Sovacool, the term prosumer thus refers to persons as actively engaging agents in the energy sector responsible for the management of their very own energy production and consumption. The remaining production surplus can be fed into the grid when their energy needs are met. So, the term prosumer is mainly used to describe households, businesses, communities, organizations, and other institutions or entities taking advantage of technological innovations such as solar panels and smart meters. The combination and interconnection of these technological advancements allow electricity generation within their energy management system and provide electricity for different energy systems. This contributes to diversifying the energy mix, increasing localized EE, lowering the cost for participating agents, and reducing GHG emissions from the electricity system and private transport (Parag & Sovacool, 2016). In brief, prosumers are generating electricity for their own demand while simultaneously being able to feed electricity surpluses into the public grid. At the same time, they continue to be able to consume electricity from the public grid in case their own PV panels are underperforming (von Arnim & von Arnim, 2020, p.799). As pointed out by Li:

> "(…) this rise in user participation blurs the line between production and consumption activities, with the consumer becoming a prosumer. The idea

> of prosumption, i.e. that production and consumption take place at the same time, is likely to inevitably become one of the major industry trends in the near future" (Li, 2018, p.79).

So, the main benefits of prosumption lie within the active and direct participation on behalf of ordinary inhabitants within the energy trade. This empowerment of citizens to participate in the energy exchange might, in turn, create a positive feedback loop toward the acceptance of RE production and generate future investments in RE technologies. Furthermore, as pointed out by Lowitzsch and Hanke, the idea of "prosumership" displays a variety of additional benefits for both the consumer as well as the promotion of low-carbon emission technologies:

> "When consumers become prosumers of renewable energy (RE), they produce a part of the energy they consume, thus reducing their overall expenditure for energy, and the sale of excess production gives them a second source of income. These positive effects on disposable household income further increase when prosumership is coupled with energy efficiency (EE). Investing in RE while at the same time reducing consumption by improving EE reduces the amortization period of the investment, since less money is spent to buy energy and a larger share of (excess) production may be sold to the grid" (Lowitzsch & Hanke, 2019, p.6).

Economically, individuals are increasingly incorporated into an incentive structure that promotes ownership of RE resources and electricity generation while allowing new actors and investors to enter the local energy market. This, in turn, enables customized, tailor-made energy solutions. As pointed out by Schill *et al.*, such an approach is more in line with concrete consumer preferences within communities and strengthens their participation in the RE sector. This consumer engagement is likely to positively affect the acceptance of energy transitions. Additionally, electricity costs have the potential to become less volatile, and competition between energy producers is increasingly shifting to the local level. Prosumption is, therefore, potentially able to generate a higher level of flexibility within the energy system by stimulating demand-side energy management and allowing for sector coupling through which prosumers can use auto-generated energy for electric cars or other technical devices of daily use. A major advantage of prosumption is related to grid relief and avoiding distribution losses of electricity. Within this process, the prosumers' capacity to store grid energy can become vital as it increases network utilization opportunities (Schill *et al.*, 2017, pp.10–13). So, community energy storage (CES) is emerging as a decentralized solution contributing to peak shaving and, thus, consumption reductions and grid relief during periods of high energy demand. By doing so, local storage mechanisms become part of the critical infrastructure and enable the decoupling of communities from the public power grid. This reduces energy demand and electricity costs due to reinforcement deferrals (Koirala *et al.*, 2019, p.228). As pointed out by Schill *et al.*, however, prosumption is also associated with efficiency losses compared to centralized power systems due to additional costs resulting from higher balancing necessities of increased RE feed-in and load variability. To achieve a balancing effect, an energy system's larger geographical extension can benefit due to complementary time profiles stemming from different renewable and non-RE resources. This results in higher energy resource flexibility at different geographic locations. Decentralized storage through PV batteries does not display such advantages and calls for additional RE storage capacities to be installed. Consequently, further investments into smaller batteries at a higher specific cost must be included in the calculation (Schill *et al.*, 2017, p.14).

As shown until now, technological developments surrounding prosumption have primarily been created to strengthen energy security and reduce the cost of energy. Prosumption, however, goes beyond these aspects by adding a climate-related dimension (Hall *et al.*, 2020, p.22; Gährs *et al.*, 2016). Promoting prosumage shows the effect of growing percentages of energy generated from renewable resources and differentiation of the energy matrix. So, while being a viable option to combat energy poverty, prosumage can also be considered an emerging approach to curb energy sector-related CO_2 emissions. In fact, Flaute *et al.* argue that a sharp increase in prosumer households is indispensable to comply with national emission reduction targets and climate change-related international agreements. The 2015 Paris Climate Accord obliges highly industrialized nations such as Germany to reduce GHG emissions by up to

95 percent by 2050 (Flaute *et al.*, 2017, p.154). Low-carbon energy innovations are, therefore, urgently needed. In this context, IRENA supports prosumer-based energy supply and the emergence of a necessary "do-it-yourself" (DIY) behavior to more precisely adapt energy generation and consumption to distinct local settings and to make more efficient use of DER in the sustainable energy transition (IRENA, 2019, pp.10–11).

6.3. Smart Grids

IRENA perceives effective "intelligent" energy solutions "(…) to include communication, information management, and control technologies that contribute to the efficiency and flexibility of an electricity system's operation" (IRENA, 2015, p.5). According to the former chief of the German grid regulation agency (Bundesnetzagentur) Matthias Kurth, the term "smart grid" as such is an outright misnomer. The existing transmission system in Germany and elsewhere can be regarded as relatively stable and capable of managing the incorporation of large-scale electricity generated from renewables without major disruptions and blackouts. However, medium and low grids are too static and have little controllability. What is missing are smart meters and other control devices inside the grid and digital methods to measure the incorporation and fluctuation of wind and solar energy. Second, the electricity generation and consumption levels are indirectly impacting the grid. Due to the volatile production of energy from renewable resources, a more efficient and decentralized form of energy generation can lead to a higher feed-in of renewable resources into the grid. A key idea of smart grids is to shift the focus from consumption to the available energy level. Adjustments must be made on both the supply (refraining from feeding-in energy into the grid) and on the consumption side (adapting consumption to available electricity in the grid). To gain the necessary information on customer behavior, the implementation of digital devices is needed, and the analysis of obtained data needs to serve as a basis for business models contributing to reducing energy costs in the long run. Like other technological advancements, the constant data analysis needed for the effective use of smart grids has to be accepted by the general population and entails a cultural dimension (Kurth, 2013, pp.11–12). Smart grids can, therefore, be referred to as a concept to modernize the existing electricity grids by implementing ICT to make the demand-response equation more efficient, enable smart power storage, and reduce losses. So, to implement smart grids, real-time consumption data analysis is obligatory by installing telemeters, RF technologies, GPRS/UMTS mobile data, FTTH fiber optic networks, and TCP/IP internet protocol systems. Such an approach is intended to achieve more flexible pricing, payment, and billing options, provide incentives for higher competitiveness, and develop a dynamic energy market. This competitive moment can then reduce consumption due to increasing EE and grid resiliency while positively reducing GHG emissions (Nocentini *et al.*, 2013, pp.19–20).

Collecting vast amounts of data is key to effectively managing a smart grid's functioning and understanding and predicting energy production and electricity consumption patterns. To take on these challenges, Horn and Mirzatuny argue: "A climate-stable energy system (…) will be far more decentralized, integrating a myriad of distributed energy resources and managing multidirectional power flows, fast-growing penetrations of intermittent renewable generation, and massive new electric loads, including plug-in vehicles" (Horn & Mirzatuny, 2013, p.47). The concrete decision on which smart grid technology to deploy depends on the exact electricity system. This is because options to implement RE installations, digital technology, and smart grids vary substantially and must be adjusted to the distinct characteristics of the local energy matrix, geographical conditions, and local demand and consumption pattern (Marczinkowski *et al.*, 2019).

As highlighted by IRENA (2013), three aspects of smart grid configurations have to be considered before feeding in electricity generated from renewables:

1. An RE contribution to the grid of less than 15 percent does not demand the installation of smart grids.
2. With a contribution of renewables between 15 percent and 30 percent, implementing smart solutions is recommended.

3. A level above these percentages is said to demand the installation of digital grid technologies to maintain the grid's proper functioning.

The most basic forms of smart technologies are Distribution Automation (DA)[33] and Demand Response (DR).[34] The latter is being deployed if a peaking plant or energy storage facility is needed. DR[35] is economically and energy efficient regarding conservation voltage reduction (CVR). Advanced metering infrastructure (AMI) and advanced pricing can be helpful in terms of nudging consumer prices and aligning them with the real cost of production. Furthermore, Smart Inverters[36] and renewable forecasting can benefit from 10 percent to 15 percent of renewable contribution to the grid and are indispensable if renewable resources contribute by more than 30 percent to the maximal grid capacity (IRENA, 2013, pp.11–12). However, smart grids should be

---

33 According to IRENA: "DA is considered a core part of a smart grid, interacting with almost all other smart grid applications and making the grid more efficient and reliable. DA helps enable RE by dynamically adjusting distribution controls to accommodate variability, power ramping and bidirectional power flows. In addition, some smart inverters may become controllable DA assets themselves" (IRENA, 2013, p.27).

34 According to IRENA: "DR, also called DSM, refers to techniques for reducing electric system loads during times of peak electricity usage or when renewable output drops (...). The benefits of DR include avoiding the use of the most expensive bulk generation plants, avoiding construction of additional generation and transmission capacity, and avoiding brownouts and blackouts. (...) There are three general categories of DR: direct load control (DLC), voluntary load reduction, and dynamic demand" (IRENA, 2013, p.26).

35 According to Nojavan & Kazem a: "demand response program is defined as changes in electric consumption patterns of end-user clients in response to changes of electricity price over time or to incentive payments designed to decrease high electricity usage at high wholesale market prices times or when the system reliability problems occur. In other words, the procedure through which consumers respond to the price signals inserted in tariffs by changing their consumption patterns is called the demand response programs (DRPs)" (Nojavan & Kazem, 2020, p.v).

36 With regards to „smart inverters", IRENA highlights the fact that: "RE sources have several drawbacks from an electricity grid operator's point of view. They can cause transient grid voltage fluctuations ("flicker"), steady-state grid voltage problems and frequency deviations. However, when smart inverters are used to interface RE sources with the electric grid, these problems can be mitigated (...)" (IRENA, 2013, p.30).

applied based on a prior cost-benefit analysis (CBA) to avoid unnecessary public and private expenditures and the misallocation of resources. Within this process, transmission and distribution (T&D) systems are of vital importance as this sector shall be granted around 50 percent of the overall expected global investments in the energy field until 2035 (IRENA, 2015, p.5). To increase system flexibility, reduce costs for network reinforcement, and to create planning coherence, Distribution System Operators (DSOs) and Transmission System Operators (TSOs) have to interact closely to manage and control real-time market activities (IRENA, 2020, p.13).

6.4. Distributed Energy Resources (DER)

As these explanations have shown until now, the shift toward an increased feed-in of renewables comes with the necessity to reorganize the T&D schemes, as transmission grids account for the highest expenditures in the energy sector. While the concept of the economy of scale—according to which increasingly larger energy production units have incrementally lower unit costs—has led to the creation of even more potent turbines and power plants, their (often) geographic location away from major economic centers or populated areas tend to minimize these benefits. High-voltage power line networks and transforming stations have shown to be too limited in their capacity to include electricity surpluses and suffer from congestion and subsequent air pollution. Instead, the learning curve within the RE and in the small-scale electric power sector displayed reductions within their unit electricity production vs. unit cost ratio, while savings resulting from plants for mass power generation reached their maximum in the 1960s and have been declining ever since. Tailor-made energy solutions for final consumers were then enabled by the reduction in material and design costs for PV panels, micro-turbines, digital command, and control services, and remote monitoring (Federal Energy Regulatory Commission, 2007, pp.1-1,1-2).

Consequently, energy transitions continuously focus on geographically concentrated smart grids as a key strategy to increase EE, grid reliability, and GHG emission reductions. According to

Guo *et al.*, DERs are being used to increase the penetration of renewable sources in power distribution networks. On a technical level, DERs generally refer to various small-scale and modular devices constructed to generate electricity. In their most common form, they have been deployed as solar panels, wind turbines, towers, batteries, and other distributed storage units (on a thermal basis). Furthermore, so-called 'controllable loads' are being used to decentralize power distribution. These can be electric cars, air conditioning for stationary usage, and smart appliances. So, the major shift lies in the focus on the consumer by enabling them to become active agents within the energy production and consumption cycle. As shown, so-called prosumers became an integral component within this decentralized process. The switch toward a demand-side-driven approach has improved the energy system's reliability and carbon emissions statistics. Through the co-generation of CHP, waste heat can be used during the conversion of power to upgrade the efficiency of the distribution system. This has proven to additionally contribute to the general efficiency of smart grid energy T&D (Guo *et al.*, 2017, p.1). Adefarati and Bansal argue that DERs contribute to the balancing of the power grid while reducing the electricity injection into the power transmission lines. Power congestion (a tremendous problem especially caused by the volatile increase of electricity generated from renewable resources) is therefore likely to be reduced. On the contrary, implementing a variety of integrated DERs enhances the stability of power flows within the grid and the power supply. The latter is particularly important, especially during high consumption periods throughout the day. A key challenge within this context is related to backup generation options. The volatile character of renewables results in subsequent uncertainties regarding the exact power generation volumes. DERs can be installed to level out the differences in energy demand and offer and can then be deployed to minimize peak time demand. The previously mentioned peak shaving applications maximize a network's generation options while reducing the cost related to electricity load shedding.

Furthermore, DER applications can increase an optimal load demand during power supply interruptions and add electricity load supply during power cuts until normal supply is restored. Therefore, combining non-renewable and renewable DERs is recommended to reduce costs and guarantee a stable power supply. DERs can, therefore, be considered technologies for power-generating and storing purposes. Furthermore, DERs can be adapted according to demand and provide power and heating solutions for residential areas, commercial complexes, and the industrial sector. Due to the large options to apply DERs, these technologies have been used as an emergency power tool for load curtailment to minimize power demand, especially during peak hours, for CHP solutions and demand response (Adefarati & Bansal, 2019, pp.30–36). According to the European Commission, DERs allow for this wide array of application options as they "(…) consist of small- to medium-scale resources that are connected mainly to the lower voltage levels (distribution grids) of the system or near the end users" (European Commission, 2015, p.18).

6.5. Blockchain for renewable energy transitions

In this context, blockchain technology has gained prominence with regard to the expansion of RE transitions (Shafie-khah, 2020; Andoni *et al.*, 2019; Orsini *et al.*, 2019). Blockchain technology is a digital innovation that allows technologies such as DERs and the components of smart grid infrastructure to operate more effectively and efficiently (IRENA, 2019, p.28).[37] While having primarily caught the attention of the Bitcoin and Fintech industries, blockchain has either been referred to as disruptive or regarded as a short-term marketable hype (Gough *et al.*, 2020, p.5; Meinel *et al.*, 2018, p.7). The

[37] It has to be highlighted those transactions (understood as the transmission of data to be associated with values and assets) are structured in blocks, which then can extend and construct a chain. It is therefore a concept of data management embedded within DTL. The latter has been referred to as the underlying technology providing solutions to generate a consensus between all participating nodes of a DTL. And although nodes can try to incorporate potentially false data (byzantine error), DTLs can detect them by applying game-theoretical concepts. So, the decentrality prevents dangers known to occur in centralized infrastructures from manifesting. Unlike in centralized systems therefore, in which errors usually affect all other sub-services, the very dependency on a central authority does not apply to DTLs (Sunyaev, 2019, p.17).

technology as such can be defined as a "global distributed ledger, which facilitates the movement of assets across the world in seconds, with only a minimal transaction fee. These assets can be any type of value, as long as they can be represented digitally" (Frøystad & Holm, 2015, p.8).

A key element within this context is the so-called Distributed Autonomous Organization (DAO).[38] This specific decentralized organizational form is carried out on a blockchain basis and "enables people to coordinate and govern themselves mediated by a set of self-executing rules deployed on a public blockchain (…)" (Hassan & De Filippi, 2021, p.1). DAOs enable the self-coordination and self-governance of a group of participants acting in favor of a commonly defined goal. The DAO source code is deployed in the blockchain and can issue smart contracts. The smart contract code, in turn, regulates the rules of engagement between the participants. So, the main organizing principle lies in sharp contrast to traditional and hierarchically structured organizations in relocating decision-making rights to the network participants, specified interaction rules, self-executing smart contracts, and independence from any form of centralized control (ibidem, p.4).[39]

---

38 Interview conducted with Juan Ignacio Ibañez (Centre for Blockchain Technologies, University College London) on September 29th, 2020.

39 To express it in another way, DAOs are business models that differentiate themselves from traditional businesses or companies by relying on two fundamental elements: *smart contracts* and *blockchain*. While DAOs may have the same structure (departments, functions, etc.) as a traditional company, they differ in the way in which the relationships between parties are established. They are not based on decisions made by individuals, but rather on self-executing contracts (smart contracts) programmed on the basis of blockchain protocols. In other words, DAOs ensure that relationships (e.g., fulfilment of contractual clauses, achievement of objectives, etc.) between individuals that form part of a company or business do not depend on the perceptions or decisions of other people, but are based on objective indicators. Those indicators are included in smart contracts, which operate on digital "if-then" statements and are programmed on the basis of blockchain protocols to guarantee the security of transactions. When, within this context a certain pre-programmed condition (objective indicator) is triggered, the smart contract, automatically executes the corresponding contractual clause without human intervention. Due to this logic, DAO's are said to bear a disruptive potential with regard to business management logics and to alter hierarchical top-down structures (Wang *et al.*, 2019).

Blockchain's main objective is to increase trust[40] within business interactions.[41] So, the management of processes and databases, which have traditionally followed centralistic approaches regulated by and emanating from appointed authorities, and a centralized synchronization of processes have been prone to certain risks. The latter manifest themselves in performance deficits, reduced system stability, losses of authenticities, and internal and external integrity (Mika & Goudz, 2020, p.37). According to Casey and Vigna, these defects are largely a result of the nature and interest of middle-managing institutions. And even the present-day organization of the globalized economy is primarily based on a "centralized trust model", in which governmental and non-governmental institutions monitor and record our daily transactions and value exchanges. They are exclusively entrusted with updating the information obtained within the ledgers they control. The knowledge gathered on customers provides them with unique power potential

---

40 As throughout most social fields, using the term "trust" within the context of blockchain remains a challenging task. The latter results first and foremost from the fact that no consensus regarding a common definition of the notion exists. Cognitive processes of decision-making under uncertainty on whether to trust or to distrust are characterized by a high level of complexity (Arduin, 2021). And although it is argued that trust is quantifiable and therefore measurable (Grüner *et al.*, 2018), the very complexity of blockchain governance reduces the likelihood of considering all relevant variables for its measurement. So, focusing on increasing trust or decreasing distrust remains empirically challenging. Confidence has therefore been introduced as a concept which does refer to the existence of expectations regarding a person's actions or a systems' functioning, but rules out considerations of risk, vulnerability and uncertainty. Instead, increasing confidence looks at enhancing predictability (De Filippi *et al.*, 2020, p.11).

41 This includes the strengthening of *honesty* as both an ethical and economic element within business relations, the consideration of *fair exchanges and goodwill* (in spite of diverging interests), *accountability* by keeping commitments and maintaining compliance toward investors, shareholders, employees and customers alike, and, finally, to *eliminate distrust* through a general policy of openness. Overcoming distrust and creating transparency is the central element in carrying out global business relations. While intermediate organizations have historically overtaken the execution of transactions—from banks, to credit card firms, to major energy companies and electricity distributers, etc.—and have therefore been assigned (and often failed) to deliver upon the above-mentioned premises, in a blockchain dominated world trust results from the functioning of the network and/or objects within the network (Tapscott & Tapscott, 2016, pp.55–58).

to decide who will be able to participate or who will get excluded from the right (and/or the extent) to participate in commercial relations (Casey & Vigna, 2018, p.87). So, the currently ongoing power shift toward a decentralized and increasingly digitalized economic ecosystem, in which owners of internet-connected technical communication devices are replacing centralized verification and authorizing institutions, needs to be accompanied by mechanisms validating that the participating nodes operate in the interest of the common good. Generating consensus on multiple issues among millions of potential computer users will likely impact corporate and political governance (Derbali *et al.*, 2019; Casey & Vigna, 2018, p.111).[42] It must be pointed out that blockchains are based on different validation methods. While there are a variety of different validation mechanisms, those most commonly referred to are PoW as in the case of Bitcoin, "Ripple" as a distributed consensus, and PoS as used in the case of the Ethereum platform.[43] The common idea of all these validation methods is based on correct transaction validation while ruling out manipulation attempts. Finally, *block chaining* refers to the process where the new block, which has been validated, is being chained into the blockchain. The updated version of the ledger is being transmitted to the network. The latter

---

42 According to Frøystad and Holm, the blockchain process as such is subdivided into *five* distinct features. First of all, the *transaction definition* takes place, where the sender starts a transaction and advises the network about it. Within this process, the receiver's public address, the exact transaction value and a transactions' authenticity through a cryptographic signature are being provided. In a second step, the *transaction authenticity* takes place between the networks' nodes. The participating nodes are being send an information message and have to authenticate its validity by decrypting the digital signature. The transaction is then being authenticated and inserted into a pending pool of transactions. In a third step, the *block creation* takes place. The pending transactions are being inserted in an updated ledger—also referred to as block—by one of the participating network nodes. At a distinct time-interval, the node communicates this new block to the network, where it needs to be validated. This fourth step is called *block validation* and occurs as soon as validator nodes are being send the transmitted block to verify the correctness through an iterative process. This process needs the consensus of the majority of the network nodes (Frøystad & Holm, 2015, p.11).

43 For all distributed consensus algorithms with regards to blockchain application to the field of energy see: Andoni et al. (2019, pp.147–150).

process takes three to ten seconds to complete (Frøystad & Holm, 2015, p.11).

6.6. Smart contracts

To successfully implement blockchain in general and in the energy sector in particular, so-called smart contracts need to be applied. According to Di Silvestre *et al.*, a smart contract can be defined as a "(…) contract, suitably coded, which automatically verifies the occurrence of certain pre-defined conditions and executes actions when the conditions between the parties are fulfilled and verified" (Di Silvestre *et al.*, 2018, p.1). In a similar fashion, Andoni *et al.* argue that the value of blockchain as such can, in fact, only be unfolded if combined with smart contracts, which they regard as:

> "(…) user-defined programs that determine the rules of writing in the ledger. Smart contracts are executable programs that make changes in the ledger and can be triggered automatically if a certain condition is met, such as if an agreement between the transacting parties is honoured. Contract terms are recorded in computer language encoding legal constraints and terms of agreement. Smart contracts are self-enforceable and tamper-proof bringing about significant benefits such as removing the intermediaries and reducing transacting, contracting, enforcement and compliance costs. An additional benefit is that low-value transactions can be made cost-efficient, while blockchains can ensure interoperability between transaction systems" (Andoni *et al.*, 2019, p.146).

Going beyond these characteristics, Kirli *et al.* attribute trustworthiness, transparency, accessibility, high-level security, speed and reliability, and self-verification of the smart contract codes (by smart contract languages and the blockchain) as key elements of smart contracts. Due to these traits, smart contracts reduce transaction costs and minimize time horizons to carry out operations. Additionally, computational performance and expense are described as distinct smart contract features. Participating nodes must contribute some of their computational power to execute the contract code in the virtual space. This resulting transaction cost obliges the owner of the smart contract. On the Ethereum platform, gas is the unit of measurement in the virtual space and is entitled the Ethereum Virtual Machine (EVM). As with Bitcoin, the invested energy intensity increases relative to the increase in complexity of carrying out the operation. In public blockchains, the cost of gas results from the demand and supply. Miners can offer the power of their computers to many agents. This leads to gas price fluctuation between 10 Gwei to 100 Gwei in a year while peaking above 400 Gwei during high congestion periods. In private blockchains, no specific financial incentive exists to participate in the execution process, and reasons to participate in this process are attributed to other advantages like cheaper and renewable-based energy supply. The smart contract logic is grounded in the idea that the initialization of the contract starts by commanding to read the generator's capacity and price offer. Consumers are then communicated the offer, and the bidding process is open. The clearing price mechanisms vary but are mainly carried out on a double-auction basis. The latter ranks bids and offers. What follows is a grid power flow assessment, an update of the smart contract, and the verification of the energy transaction by using the generator's recordings of the smart meter. Total units and generation length are verified, and scaling or penalties are added. The clearing is then authorized, and the transaction is irreversibly registered (Kirli *et al.*, 2022, pp.5–6).[44]

However, as pointed out by Gough *et al.*, the reference to a "legal contract" is misleading as smart contracts do not consist of the traditional elements to be legally binding (Gough *et al.*, 2020, p.13).

6.7. Processes of blockchain classification

Blockchains can take on a variety of different forms. Differences can be made between the data access options and how consensus is generated within the block's validation process. In so-called *public blockchains*, everyone is to be granted the right to access the network

---

[44] Additionally, Kirli *et al.* argue that there are six different layers which smart contracts comprise of. These are: 1) information from agents, on devices applied and information from the grid, 2) consensus and control algorithms for energy management, 3) smart contracting managing the financial and gas transactions, 4) verification and encryption, 5) computational processes, and 6) communication of the physical transfer of the information between the members (Kirli *et al.*, 2022, p.7).

and to participate in the validation of the transaction. Public blockchains are based on a variety of consensus. However, the most prominent ones remain PoW and PoS-based consensus algorithms (Mika & Goudz, 2020, p.50). In these so-called *permissionless blockchains,* each participating node can write and enter information into the blockchain. *Permissioned blockchains,* on the other hand, demand access permits for nodes to take part in the ledger's validation. This is usually being done to generate a higher level of trust (Gough *et al.*, 2020, p.14). Participating nodes do not know each other within permissionless blockchains, and "(...) trust emerges from game-theoretical incentives" (Mattila, 2016, p.8). According to Andoni *et al.*, this very incentive structure within the ledger management process includes the deterrence of selfish, non-compliant behavior through "(...) spending resources such as computational work, electricity or penalization (...)" (Andoni *et al.*, 2019, p.147). Permissioned blockchains are usually private and only grant access to selected people. Access rights are being decided upon and granted by different institutions. A main difference between public and private blockchains is furthermore reflected in the level of decentralization and the level of anonymity provided. Legal entities usually operate private blockchains (Mika & Goudz, 2020, p.51). The incentive structure to maintaining compliant behavior and cooperation is based upon the participatory nodes' known identities. The latter structure rules out artificial incentive schemes while network operations can be accelerated, more flexible, and more efficient. On the downside, however, this comes at the expense of a lower level of security, a reduced scope of immutability, and resistance to avoiding censorship (Mattila, 2016, p.8).

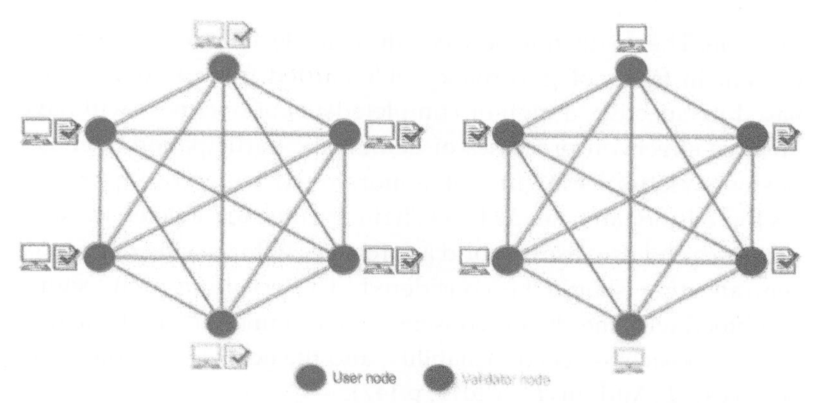

Figure 2: Graphic taken from Andoni *et al.* (2019, p.147) "Classification of blockchain architectures, public permissionless ledgers."

A third, intermediate, and consensus-based type also exists, which displays a combination of public and private elements. A variety of authorized entities or organizations can operate these *consortium blockchains*. The fact that their functions are only open to its members gives them more flexibility for business applications. Furthermore, they are said to be more efficient because using the energy-intensive PoW consensus is unnecessary (Meng *et al.*, 2021, p.1369). While potentially also expandable to the public and administrative sector, the high control potential of this blockchain type due to limited access options is especially preferred by companies within the energy sector. Consortium networks usually pay high attention to data protection, which comes at the cost of lower transparency and supervision options. The latter, however, have originally been at the very heart of blockchain innovation (Mika & Goudz, 2020, p.51). According to Zheng *et al.*, the read permission of blockchain transactions and, therefore, the decision on whether the information is publicly available or not depends on the consortium or organization running the very blockchain within both private and consortium blockchains (Zheng *et al.*, 2018, pp.357–358).

Andoni *et al.* concretize these three blockchain types by distinguishing between a *general* and *specific purpose* and *open-source* and *closed-source* blockchain. Platforms such as Ethereum are general-purpose blockchains created to allow various applications to function. The distinction between the exact level of openness is important in terms of governance. If regarded from a community-based perspective, a high or complete level of openness will have positive effects on processes of democratic participation, whereas business platforms might seek a more restricted access approach. As there is no tailor-made blockchain architecture able to serve all purposes and use cases, hybrid approaches with varying degrees of centralization should be considered. The resulting architecture, combined with the chosen consensus, determines key performance indicators such as speed, scalability, and the necessary resources to be invested (Andoni *et al.*, 2019, p.147).

6.8. Limitations of blockchain

As with most innovations in their initial stages, the novelty of blockchain also displays certain limitations regarding its applicability to the energy sector.

As highlighted by Gough *et al.*, an operating network of electricity is considered a natural monopoly, which comes with large market entry expenditures such as capital cost, economies of scale, or economies of scope. This, however, provides selected players "first mover" advantages while limiting possible market access options for others (Gough *et al.*, 2020, p.16). So, a major challenge represents the centralized structure characterizing most energy markets. Few key operators are benefiting from economies of scale and economies of scope. Consequently, blockchain in the energy sector could suffer from reduced scaling opportunities within an economic environment averse to competition and changes to their core business models. With regard to the technological side, the trackability of electricity once fed into the grid remains a main challenge (Gough *et al.*, 2020, p.19). Furthermore, the so-called 'scalability trilemma' (Gough *et al.*, 2020, p.19; Zhou *et al.*, 2020, p.16440, Hafid *et al.*, 2020, p.125244) is highlighted in the academic discourse. Permissionless blockchains allow for the creation of an ecosystem by incorporating vast amounts of nodes. However, this comes at the expense of transaction speed and high electricity costs due to the necessity of using the PoW consensus mechanisms. As a result, it is challenging to guarantee decentralization, scalability, and security simultaneously. In most cases, a maximum of two of these three benefits may result from blockchain application to the energy sector under the current state of technological development (Gough *et al.*, 2020, p.19).

Additionally, security concerns are raised around the issue of a necessary blockchain size before the network is big enough to verify all transactions. Given that the energy system is part of the critical infrastructure in most countries, possible attacks or manipulations of blockchains can potentially cause large-scale damage and have negative societal impacts. On the other hand, rapid blockchain growth with millions of DERs and high transaction volumes might induce technical challenges. Storing a copy of all daily transactions might lead to technological difficulties. Furthermore, technological challenges related to transaction sizes and speed must be improved. While Ethereum platforms were processing around seven transactions each second in 2019, traditional credit card operators, in comparison, were handling up to 5,000 transactions per second (Gough *et al.*, 2020, pp.19–20). This number has risen to around 30 but is expected to allow 100,000 transactions per second on the Ethereum 2.0 platform (Worldcoin, 22.03.2023).

From a technical perspective, blockchains lack standards for so-called application programming interfaces (APIs), which can lead to difficulties regarding interoperability with different technologies. Additionally, private key management might cause problems if a private key gets lost. The latter key verifies a participant's identity; if it disappears, access to a blockchain's digital assets is impossible (Teufel, 2019, p.7).

Additionally, most blockchain energy projects run on Ethereum, which has received most cyber-attacks in the past. Even minor programming errors within smart contracts (as an integral element for blockchain energy trading) can lead to their manipulation. Furthermore, if peer-to-peer networks break down, the supply cannot be secured anymore. It then remains highly difficult to detect those responsible for these failures. Cryptographic algorithms are central to a blockchain's functioning to ensure the blockchain operates securely. However, unknown vulnerabilities within these algorithms might not be rectifiable (Teufel *et al.*, 2019, p.7).

Furthermore, legal accountability issues might arise due to the non-existence of hierarchies within blockchains. For example, ownership rights regarding the same asset products traded simultaneously might lead to legal challenges (Gough *et al.*, 2020, pp.19–20). As Teufel *et al.* and Gough *et al.* point out, data protection is of specific concern within this context. Based on the current European Union's framework regarding its General Data Protection Regulation (GDPR), these protection rights are not sufficiently guaranteed on a blockchain basis. This, however, is important as all data within permissionless blockchains are visible to all nodes (Teufel *et al.*, 2019, p.7; Gough, 2020, pp.22–23). These deficiencies are recognized by the EPRS. As one natural or legal person acting as the data controller is being replaced by different participating nodes, ascribing accountability and responsibility remains a (both legally and practically) difficult task. Due to the European case law developments, defining the qualification for (joint) controllers is difficult.

Furthermore, the GDPR framework (Articles 16 and 17 GDPR) highlights that data should be modifiable and erasable. Blockchain, however, is based on making modifications impossible to generate trust. Consequently, it is subject to debate whether DLTs are complying with Article 17 GDPR. Therefore, private and permissioned blockchains are more likely to comply with these legislations than permissionless blockchains (EPRS, 2019, p.101).[45]

6.9. Conclusion

This chapter has shown that applying blockchain to the energy sector displays various benefits and can potentially contribute to the global energy sector's overall transformation.

---

[45] Considering these legal uncertainties, a general question raised by Gough *et al.* is whether a blockchain application is really needed if hierarchies within smaller social systems with centralized databases and high level of trust among participants exist and if these databases are not underperforming those of decentralized ledgers (Gough *et al.*, 2020, p.19)

First and foremost, this is because DTL technology can potentially contribute to taking on the challenges of balancing supply and demand more effectively. Given the volatile generation of RE on the one hand and the lack of storage capacities on the other hand, reshifting the focus toward smart grids represents a viable solution to increase coordination between energy generation and consumption. This potentially leads to positive effects with regard to more efficient use of RE and preventing electricity losses in the grid. If so, the balancing effect allows for increased final consumption of renewable sources (and not only their percental increase within the primary energy matrix), thereby reducing the necessity for fossil fuel usage.

Second, as within other areas where blockchain has been applied, this technological innovation contributes to the decentralization, transparency creation, democratization, and strengthening of consumer participation in one of the key economic fields. Blockchain allows prosumers to participate more effectively in the energy trade by enabling them to sell their self-produced (mainly solar) energy. Due to their increasing numbers, individuals are increasingly empowered to participate and gain importance as energy producers and traders. In contrast, the traditional characteristics of the energy market, with its highly centralized structure and a few selected operators, continue to erode.

Third, the beforementioned balancing effect comes with the possibility of reducing GHG emissions. Applying smart grids on a blockchain basis can be considered a viable bottom-up strategy—as pointed out within the international normative framework analysis—in the fight against climate change in general and against CO_2 emissions in particular. However, the latter is still in its initial stages, and it remains to be seen how far blockchain infrastructure components shall undergo a process of innovation and gain broad consumer acceptance to tackle the existing disadvantages.

These difficulties are predominately of technical, legal, and security-related nature. Regarding the technical components, most blockchain platforms run on energy-intensive Ethereum and PoW consensuses. While allowing for the participation of large numbers of nodes and the strengthening of the validation process, CO_2 reductions are still difficult to achieve and depend on nearly perfect configurations of all technical components involved. There is no blueprint since every locality, and its characteristics have to be adapted to with regard to the technical configuration. Second, legal and accountability issues related to blockchain need to be resolved to allow for the generation of trust and stronger involvement of the business sector in blockchain's further development. Finally, security-related concerns exist around the possibility of manipulating smart contracts and blockchain algorithms with damaging consequences for the energy system as part of the overall critical infrastructure. While these latter considerations are at the core of blockchain's "scalability trilemma", this chapter has shown that the energy sector (due to its technical complexity and data-intensive nature) remains one of the most promising fields for blockchain application.

7. First Empirical Case Study: The Brooklyn Microgrid (BMG)

7.1. Introduction

To approach this article empirically, the BMG, the first comprehensive project to enable community-based RE trading on a blockchain basis, shall be analyzed. So far, the BMG has primarily been subject to research focusing on its functionality from an (industrial) engineering perspective (Orsini *et al.*, 2019; Andoni *et al.*, 2019; Mengelkamp *et al.*, 2018). However, before diving deeper into the Brooklyn case, blockchain-based microgrids' particularities and features deserve a more detailed analysis. This is because although microgrids are integral to the development of smart grids, both concepts have to be considered in a differentiated manner in as much as "smart grids take place at larger utility level such as large T&D lines, microgrids are smaller scale and can operate independently from the larger utility grid" (Yoldaş *et al.*, 2017, p.206).

In a first approximation, Hatziargyriou *et al.* define microgrids as those which:

> "(...) comprise Low Voltage distribution systems with distributed energy sources, such as micro-turbines, fuel cells, PVs, etc., together with storage devices, i.e. flywheels, energy capacitors and batteries, and controllable loads, offering considerable control capabilities over the network operation. These systems are interconnected to the Medium Voltage Distribution network, but they can also be operated isolated from the main grid, in case of faults in the upstream network. From the customer point of view, microgrids provide both thermal and electricity needs, and in addition enhance local reliability, reduce emissions, improve power quality by supporting voltage and reducing voltage dips, and potentially lower costs of energy supply" (Hatziargyriou *et al.*, 2006, pp.1, 2).

According to Schwaegerl and Tao, a microgrid is a platform that integrates supply, storage, and demand resources in a distinctly localized distribution grid. The local supply of electricity within microgrids stems from nearby loads on an LV level basis, which come with microgeneration capacities operating below megawatt levels. Second, microgrids should be able to connect to the public grid and operate on an islanded basis. The latter modus (as in the case of the BMG) was designed in the first place for emergencies and possible interruptions of the public grid. To conduct such operations on a microgrid basis and to guarantee electricity transmission stability, the availability of sufficient microgenerators and a large capacity of storage mechanisms have to be put in place. If guaranteed, the main innovation of microgrids lies in the combination and management of small-scale generation, load controlling, and the regulation of emissions. As such, economic, technical, and environmental considerations overlap and enter into the microgrids' design. By doing so, possible conflict potential due to multiple and diverging interests of local producers, large-scale electricity distributors, and consumers regarding the grid energy matrix can be reduced significantly in advance (Schwaegerl & Tao, 2014, pp.4–5).

Introducing blockchain to microgrids results from the necessity to better coordinate and optimize demand and supply on bi-directional, demand-responsive energy transmission and peer-to-peer basis. By deploying such an infrastructure, advancing energy technology became a tool to guarantee energy security during power outages. For example, Superstorm Sandy swept across 24 US states in late October 2012, mainly impacting New Jersey, Connecticut, and the New York area. The combination of rain, strong winds, storm surges, and flooding caused large-scale property damage and affected power plants, substations, petroleum refineries, and terminals. As a result, electricity disruptions and longer periods to restore grid connections and supply chains exposed the vulnerability of the public grid system in the United States (US Department of Energy, 2013, pp.2–8).

Against this backdrop, the rethinking of energy exchange paradigms as developed by the founders of LO3 Energy gained ground. The company created the Exergy system and thereby:

> "(…) a technology platform that is layered on top of the existing utility grid and enables rapid adoption and integration of distributed energy resources (DER), which include storage, solar, and other renewables. Through Exergy, consumers and prosumers can transact the value of their DERs. Exergy also enables the connection of consumer/prosumers to grid operators that need to alleviate a specific grid issue" (Orsini *et al.*, 2019, p.224).

As this statement highlights, the LO3 Energy project is trying to design solutions to manage the tremendous complexities resulting from RES power production and consumption in the wake of the growing inclusion of RES into the grid.

According to Nieße *et al.*, the power system transformation demands the recognition and management of distinct individually designed, configured, and redistributed loads of small-scale RES and the volatility of RES production due to place-specific climatic and meteorological conditions. Based on their market-oriented approach, the optimization of supply and demand by different coalitions needs to be accompanied by the implementation of ancillary services for grid stabilization (Nieße *et al.*, 2012, p.1). Lv and Ai point in a similar direction when arguing that a dynamic energy management strategy on a microgrid-based active distribution system (ADS) needs to be applied to adapt to the large-scale inclusion of RES, to smooth the power flow, reduce losses of power, and better manage the voltage profile in active distributive networks (ADN). Such an approach could then be expanded to hybrid energy distribution networks (Lv & Ai, 2016, pp.421–422).

7.2. Blockchain-based microgrid components

Based on the technical details outlined until now, an analysis will be undertaken of Mengelkamp *et al.* (2018) regarding the seven necessary components of blockchain-based energy markets to see how far they have been applied to the BMG. According to the authors, these components are:

1. The microgrid's clear objective must be defined and entered into the grid's technical design. Depending on the state-of-the-art available energy generation, distribution, and trading technology, the individual objectives can be conflicting and/or contradicting. The primary goal of microgrids should be to give all community inhabitants access to these grids, to generate a stable energy supply (even if the main grid supply is interrupted), and to reduce environmental impacts through the usage of solar or other (renewable) energy sources.

2. The entry and connection junctures to the micro- and subordinated grid are important for balancing energy supply and demand. At these junctures, the exact energy level can be measured through e-metering technologies, thereby contributing to regulating energy flows in the grid. Within this context, the distinction between physical and virtual grids is important. Their difference lies in their real level of connectivity and the options to disconnect from the microgrid (physical microgrid) versus the linkage of microgrid participants via an information system platform (virtual grid).
3. The information system monitoring the market activities needs to ensure equal access to all market participants in a non-discriminatory way. As mentioned before, blockchain-based protocols enable software applications and smart contracts. Smart meters are necessary to write the exact information on energy generation and demand into the blockchains. The exact mechanism to verify the transactions depends on the microgrid's layout. Single identity-based consensus mechanisms, which allow for a hash-based authentication of each user, should be used. By applying these mechanisms to a microgrid market, no user can register with multiple identities. Instead, identities can be verified by a central, that is, a government entity, before giving access to the market. The re-verification through the chosen consensus mechanism is then based on the assigned identities.
4. The information system implements the distinct market mechanism, whose main objective lies in the correct processing and handling of the buy-and-sell orders on a real-time basis while canalizing the minimal and maximal possible bidding options related to energy quantities. The market mechanism design should reference differences regarding the market time horizon structuring. Regarding local microgrids, intraday trading and the closed order book (which is to be cleared at distinct time intervals) are considered preferable.

5. Based on the market mechanism chosen, the pricing method has to be applied to regulate the energy exchange in its most efficient form. As outlined before, localized microgrid energy exchange comes at a differentiated cost compared to auctions. Price signals—high or low—will be indicators of available energy supply and demand. The microgrid can (from an economic perspective) be profitable if prices are lower than those charged by main grid operators. Higher prices can be demanded only if awareness regarding the socioeconomic importance of localized RE consumption and the community's consciousness on environmentally friendly energy generation.
6. An energy management trading system (EMTS) has to measure demand and supply data in real time. Considering the latter, the EMTS needs to construct its specific bidding design, thereby paying reference to the estimated quantities of demand and available supply. Rational market actors look for hours to sell their energy at a maximum price while reducing individual energy expenditure. However, EMTS will buy energy when prices drop below the maximum level. To create an efficient energy trading system and to foster the beforementioned socioeconomic factors, the EMTS system needs to be granted full access to the blockchain of those participating in the local energy trade.
7. Finally, legal regulations determine the operative options of microgrids and their correct insertion into an overall energy policy scheme. These regulations also determine how taxes, fees, and contributions will be charged. In other words, governmental decision-making is at the forefront of promoting or limiting the implementation of microgrids (Mengelkamp *et al.*, 2018, pp.873–875).

Consequently, the integration of an efficient information system, a market mechanism, and a price-setting mechanism represent the basis for the functioning of an efficient microgrid energy system. They should be automated through an EMTS system, which will also allow for a reduction of trading periods.

7.3. The Brooklyn Microgrid (BMG)

Based upon the features mentioned earlier, it shall now be explored how far these components have been applied to the Brooklyn Microgrid, including additional literature than those selected by Mengelkamp *et al.* (2018).

Based within the Brooklyn neighborhood of Park Slope—a neighborhood of 67,649 (New York City Neighborhood Tabulation Areas 2010)—the community microgrid has gained prominence as the world's first comprehensive project. According to Orsini *et al.*, Park Slope was chosen due to the preexistence of private consumer investments in solar energy (private rooftop solar panel installments) and an expressed community interest in installing a microgrid in the neighborhood. In other words, the preexisting distribution infrastructure consisting of traditional and DERs was key in selecting Park Slope. The latter also consists of vanadium flow batteries[46]. Breaking through the difficulties of energy storage deriving from RE sources as described before, these batteries have shown to have almost unlimited storage and redistribution capacities, making them especially adaptable to the specifics of microgrids (Orsini *et al.*, 2019, p. 230).

According to Mengelkamp *et al.* (2018, p.876), the main feature of the energy grid is the *virtual community energy market platform* providing the technical infrastructure for energy trading. More specifically, the private blockchain applied is based upon the usage of a so-called Tendermint protocol. According to Di Silvestre *et al.*, the Tendermint protocol:

> "(...) is a blockchain that allows the secure and consistent replication of an application on different machines. Safe because it works even if 1/3 of the machines fail arbitrarily (capacity known as Byzantine-BTF fault tolerance) and consistent because each machine sees the same transactions and are in the same state. The advantage of using Tendermint compared to the other blockchains, is the possibility of using any programming language to write the code of the App to process the transactions to be included in the

46 According to Cunha *et al.* (2014), the vanadium redox flow batteries were developed to store large volumes of electric energy, making them especially suitable for the storage of intermittent renewable energy resources, and permitting adjustments to real time energy demand (Cunha *et al.*, 2014, p.910).

> blockchain blocks. Tendermint consists of two main technical components: a consent engine (Tendermint Core) and an interface application (Application BlockChain Interface - ABCI). Tendermint Core ensures that the same transactions are recorded on each node, while the interface application allows to communicate with Tendermint Core by processing transactions in any programming language" (Di Silvestre *et al.*, 2018, p.3).

The TransActive Grid blockchain and smart meters exist beside an analogous metering system. *The physical microgrid*, which currently incorporates ten-by-ten housing block areas, can operate in island mode. In case of power cuts, entities belonging to the critical infrastructure receive energy at fixed prices, while Park Slope residents and business owners have to get involved in the bidding process. Con Edison Incorporated is the main energy provider to the traditional grid, and decoupling only occurs in emergencies. So, generation and consumption information is being directly passed on to the participants' blockchain account through the smart meter of the TransActive Grid. Based on this buy-and-sell order information, orders are transferred to the *market mechanism*. The latter is equipped with smart contracts to carry out the transaction. Once a buy-and-sell order has been matched, smart contracts are issued, the payment will be conducted, and a new block carrying the new transaction information will be inserted into the blockchain (Mengelkamp *et al.* 2018, p.876).

So, as pointed out by Zia *et al.* (2020), DLTs and local energy markets are the defining elements within a decentralized transactive energy system. These components uphold system reliability and control through the correct integration and configuration of DERs and prosumers in the neighborhood. Focusing on the Brooklyn case, this logic does apply to the microgrid transactive energy system. The latter can then be regarded as:

> "(...) an information and communication technology-based ecosystem that used community technologies, internet, and mobile networks-based hardware/software platform to trade energy among power producers, prosumers, and consumers. The energy trading and sharing process is achieved by determining market equilibrium at market clearing price (MCP) using real-time information of bids and offers. The energy management objectives of an MG-TES can be dynamic demand supply balance, profit maximization of power producers, reduced GHG emissions, cost minimization of MG system

> and prosumers, and congestion management among others" (Zia *et al.*, 2020, p.19411).

In line with the seven components of blockchain-based energy markets outlined by Mengelkamp *et al.* (2018), participants of the microgrid market in Brooklyn are both consumers and prosumers locally. Energy trading has been limited to the purchase and selling of electricity up until now. The connection junctures between the micro and public grid are made possible by infrastructural units such as TransActive Grid meters. The connection between the microgrid and the main public grid balances electricity supply and demand. Furthermore, it is needed for balancing purposes if local electricity availability exceeds and does not meet local demand (Mengelkamp *et al.*, 2018, p.876). According to Zia *et al.*, the BMG interconnection takes as well place via the abovementioned ICT-based ecosystem MG-TES, which is based on a utility grid electrical infrastructure to allow for energy flows (Zia *et al.*, 2020, p.19418).

As said before, the decoupling of both grids can and will be conducted in emergencies exclusively. However, unlike physical microgrids, the virtual microgrid cannot disconnect from the superordinate public grid, which leads to challenges regarding the supply of the remaining customers through the main grid. The information system monitoring the market activities, as mentioned in Point Three of the analysis developed by Mengelkamp *et al.* (2018), is then carried out by the described TransActive grid platform based on the private blockchain protocol. Bottomed on the measurement of demand and supply and additional market information data (on actual generation and consumption), the energy prices are determined and displayed to the customers. The payment transaction is conducted between microgrid participants based on established regulations included in the market mechanism (Mengelkamp *et al.*, 2018, pp.876–877).

The EMTS, as described in Point Six, carries out the trading automatically, limiting the participants' interventions to selecting energy source preferences and determining certain maximum price levels. Prosumers and consumers participating in the BMG can submit orders to buy and sell energy through the EMTS platform, which displays the exact quantity and the determined prices based on the beforementioned criteria. Currently, customers are submitting the maximum price offers for selected energy sources at which they are willing to buy (mainly RE from solar), while prosumers send out a selling bid and disclose the minimum price at which they are willing to sell their generated energy (Mengelkamp *et al.*, 2018, p.876). By doing so, customers display insights regarding their socioeconomic preferences (Zia *et al.*, 2020, p.19418). The bidding is based on what can be referred to as a merit order principle. The MCP is set when the last bidder submits their bidding order within a given time interval. Consumers not going below the clearing price are granted additional energy sources they can choose from. The latter aligns with Point Five of the important microgrid elements highlighted by Mengelkamp *et al.* (2018, pp.875–876).

To increase efficiency and take on the existing challenges regarding the intermittency of power generation resources, special attention in the BMG (and in all other blockchain-based grids) has to be paid to the design and the algorithm used for energy trading. At the BMG, a double auction mechanism has been applied to the TransActive microgrid platform for consumer transactions. The latter was developed by LO3 and Siemens (Wang *et al.*, 2017, p.2). This design has been chosen to define the local MCP. The main idea of these peer-to-peer markets was to optimize the energy costs of microgrids while expanding DERs and increasing the overall system reliability (Zia *et al.*, 2020, p.19425). The latter aligns with Point Five of the analysis by Mengelkamp *et al.* (2018, p.875). However, it must be further developed and adapted to allow for dynamic trading of RE sources. As pointed out by Wang *et al.*, this is because the BMG "has no appropriate bidding strategic model for transaction parties and direct settlement based on blockchain" (Wang *et al.*, 2017, p.2). Zia *et al.* point to the relatively small quantity of participants within the BMG as a possible reason for the lack of trading system improvements. However, they see this LEM closing existing trading gaps (Zia *et al.*, 2020, p.19418). Market mechanism (4) and pricing strategy (5) are factors that also need to be improved (Mengelkamp, 2018, p.876), especially since the participation of community consumers in the purchase of RE resources is of key importance. It is the latter, which – combined with the described technology innovations – allows for a reduction in CO_2 emissions by switching toward a higher contribution of renewable resources.

The necessity to expand within this field is also highlighted by the LO3 Energy managers:

> "Exergy enables prosumers and consumers within the local market to engage and grow together to achieve economic and environmental goals. The Brooklyn Microgrid's community-based model *provides the potential* of new revenue streams, incentivizing consumers to invest in DER and to become prosumers. It creates a circular economy -renewable, reliable, and resilient with the potential to utilize resources efficiently- within the local market, similar to the online marketplace Airbnb fostered for travelers. (…) Looking toward the future, innovators such as LO3 Energy envision a landscape of distributed energy-producing Microgrids connected to utility grids throughout a vast network. Getting there requires significant investments and prototypes of community microgrids" (Orsini *et al.*, 2019, p.236).

Learning from successful blockchain microgrid concepts created globally is thus key to contributing to effective RE distribution and GHG reduction.

The seventh and last point mentioned by Mengelkamp *et al.* (2018) refers to the legal issues regarding the installation and operation of microgrids. As pointed out by a variety of different studies, one of the main challenges concerning the proper implementation and expansion of microgrids and P2P energy trading lies within the existing legal frameworks (Zia *et al.*, 2020; Orsini *et al.*, 2019; Mengelkamp *et al.*, 2018). According to Zia *et al.*, sophisticated "regulation layers" for decentralized transactive energy systems and efficient P2P trading, which include standards, laws, governmental policies, and legal requirements promoting these innovations, are inevitable due to the following reasons:

> "The legislative rules and regulatory policies are necessary for providing [a] framework for LEM design and its integration with other electricity markets and electrical network. Moreover, taxes and surcharges policies must also be defined for TESs. Governments can also introduce such incentive schemes for MG-TESs that increase customer willingness in participation and use of local RESs for reduced GHG emissions (…)" (Zia *et al.*, 2020, p.19418).

However, P2P energy trading between at least two community residents without utility companies involved has (previously) not been possible due to a lacking adaptation of US laws (Mengelkamp *et al.*, 2018, p.879).

A similar point of view regarding the current state of LO3 Energy deployed technologies, their innovation potentials, and their current limitations because of existing regulations is being shared by Orsini *et al.* (2019). Although considering the microgrid as disruptive in the energy market, they recognize its design as synergistic with traditional grids, TSOs, and DSOs. They call for an adaptation to new business models in digital energy distribution. They will, therefore, continue to convince regulators and policymakers to shift their focus from traditional supply-side investments (such as pipelines and wire) toward policies providing incentives for EE (Orsini *et al.*, 2019, pp. 237–238). A first step in this direction was made in 2020 when the BMG was approved to conduct a pilot program to trade energy on its platform. Forty prosumers and 200 consumers were included in this trial period. Consumers were enabled to place their bids for excess energy provided by prosumers and traded through the LO3-designed Pando app. It was intended to be an open platform without a minimum trading price floor. The idea was to test the feasibility and self-sustainability of the BMG in economic terms, that is, selling excess energy on behalf of prosumers to consumers without having to resell the energy to the utility company according to net metering prices. Such an approach would evade existing regulations (due to the beforementioned limitations to allow only electric utilities or licensed retail services the right to sell energy). The approval came after community members and BMG representatives met the governor's administration (Sharma, 2019).

7.4. Conclusion

This chapter—predominantly based on Mengelkamp *et al.* (2018)—has shown that only a combination and alignment of factors encompassing the 1. technical design, 2. connection points to the main grid, 3. non-discriminatory market access for all participants, 4.

distinct market designs, 5. efficient pricing mechanisms, 6. elaborated energy management systems, and 7. proper regulations will allow for a large-scale inclusion of RE and significant reductions in GHG emissions. While the first three aspects have been completely applied, the following three elements are currently subject to improvement. By doing so, LO3 Energy is open to learning from international experiences and applying the latest and most inclusive customer-friendly options. The latter is necessary to maintain and expand support and acceptance for the microgrid concept on behalf of Park Slope community members. Until now, domestic and commercial energy consumers have warmly welcomed the concept for their backup functionality in emergencies due to the growing awareness regarding the nexus of environmental degradation and energy usage, especially among urban middle- and upper-middle-class residents in the US.

From a theoretical perspective, this chapter has shown that policy learning, policy diffusion, and technology transfer approaches apply in the first place to the Brooklyn case since knowledge regarding both legal adaptations and technological innovations is key to tackling the challenges caused by CO_2 emissions from the energy sector.

The growing internalization of a necessary shift toward localized RE consumption has yet to be accompanied by a legal framework allowing for effective energy trading within microgrids. Personnel of the Brooklyn case and other blockchain-based microgrids worldwide must exchange information on experiences gained to adjust legal and technical components better. Adapting legislation is, of course, a political issue as it challenges traditional concepts of energy distribution. However, given that the BMG has become a showcase for successful blockchain-based energy distribution despite its remaining limitations, it provides hope that policymakers recognize the benefits of microgrids concerning their positive effects on environmental contributions, employment opportunities, and energy security. So, while regulations remain the key obstacle to bringing about the microgrid's full potential in balancing the volatile energy production of renewable sources, the example also shows that community-based blockchain models already present a viable option to lower CO_2 emissions through the localized consumption of RE resources.

Furthermore, the United States Inflation Reduction Act passed in August of 2022 strongly focuses on GHG reductions through EE and a sharp increase in renewables roll-out. Tax credits to invest in RE projects before 2025 include microgrid controllers for projects up to 20 MW (The White House, 2023, p.14). This can be regarded as a first step to institutionally recognizing the importance of promoting localized energy distribution and trading projects as part of an effort to mitigate GHG emissions effectively. Therefore, being part of the national contributions to meet the Paris climate objectives, RE promotion on a microgrid basis due to the Inflation Reduction Act can be seen as a bottom-up strategy inspired by the global negotiations on climate change policies. Such an interpretation stands in the liberal tradition of climate change regimes.

As the selected literature focused mainly on economic and engineering explanations of the BMG's functioning and its possible expansion, anthropocentric (human-centered) approaches as part of green theory best explain the driving motivations to engage in activities surrounding the microgrid. Business interests and a stable and cost-efficient energy supply are the main incentives to participate in the local microgrid to improve the local population's living conditions. However, the fact that various actors—from entrepreneurs to local inhabitants—are engaged in the BMG's energy exchange at least suggests that a growing awareness of the CO_2 emissions and air pollution reduction potential exists. The notion of climate justice also results from the common and equal participation rights and the inclusion of all participants in the microgrids' decision-making process independent of their socioeconomic status. Diverse interests and motivations overlap, leading to their consideration when discussing further grid expansions. These opinions are also voiced in negotiations with policymakers regarding possible legal changes on energy trading options beyond the current scope. Deriving from this observation is, therefore, the fact that blockchain-based BMG is also very much in line with ecocentric (holistic) theories of green political thought.

8. Second Empirical Case Study: WindNODE

8.1. Introduction to the digitalization of the "Energiewende"

Another blockchain-based approach was investigated as part of the government-funded WindNODE research project in northeastern Germany. The WindNODE project sought viable solutions for Germany's energy sector, such as higher RE incorporation into the grid and possibly reduced GHG emissions to make the nation's energy transition more efficient and climate-neutral (Munzel *et al.*, 2022).

Named *Energiewende,* this mammoth project entered the global discussion on the opportunities, challenges, and limits of energy transitions. The *Energiewende* represents nothing less than the biggest challenge in modern economic history. Never before has a country, let alone a highly industrialized nation such as Germany, intended to completely switch to RE while abandoning its nuclear, coal, and gas resources (Beveridge & Kern, 2013).

To make use of these experiences, Steinbacher notes that the *Energiewende,* therefore always "had an international ambition, especially with regard to facilitating the deployment of RE in the electricity sector. Making RE technologies accessible to other countries by introducing them at scale in Germany, and thereby contributing to global climate protection efforts, has been a core theme of Germany's Energiewende since its beginnings" (Steinbacher, 2019, p.127). This represents an even more challenging task considering that all have not backed Germany's energy transition, let alone the majority of European Union member states, and that its own national energy-related climate strategy shows certain contradictions (Radtke *et al.*, 2018, p.30ff.). While coal burning represents one of the sources with the highest CO_2 emissions, Germany had only recently decided to join the Powering Past Coal Alliance (2019). Finally, the European Trading System's long-time ineffectiveness is said to have been largely induced by Germany's exceptions granted to companies in the energy and emissions-heavy industrial sector (Radtke *et al.*, 2018, p.31).

And while continuing to promote RE for its climate friendliness (if compared to other sources), Germany only recently, and under legal pressure, passed stricter emission reduction policies and declared climate neutrality to be achieved by 2045 as part of its revised climate change act (German Federal Government, 2021). Russia's invasion of Ukraine, reductions in gas delivery since mid-2022, and subsequent price spikes led to a drastic rethinking of the necessity to expand renewables at the German and European levels. The implementation of the REPowerEU program represents a direct consequence of these developments (European Commission, 2023a).

Considering these policy differences between European Union countries, more attention has to be paid to innovations, such as the ongoing digitalization of energy transitions.[47] Within this context, blockchain technology implementation represents just one element to deal with the challenges resulting from the particularities of RE generation. However, although startups and major tech companies have made large investments in blockchain in various sectors, major breakthroughs beyond Bitcoin have not yet been reached. This counts especially for the energy sector, one of the most promising fields for DLT implementation.[48] According to Bogensperger *et al.*, blockchain's development is—mainly due to parallel and competing DLT technologies and lack of standardization—more of a basic concept than a consolidated technology (Bogensperger *et al.*, 2018, p.10).[49]

The latter is also due to insufficient investments in the digitalization of the energy sector. For example, a survey report for the

---

47 Despite the geopolitical developments in Europe and its implications since 2022, these differences are unlikely to be resolved in the short or medium term. Member states maintain the authority to select their nation's energy matrix according to article 194 of the treaty on the functioning of the European Union (TFEU) while changes depend on the principle of the unanimous vote in the European Council according to article 15 of the EU treaty (Fischer, 2014, p.2)

48 Interview conducted with the technical lead of BC prototype on July 28th, 2020 (via an online communication platform).

49 Other prominent approaches represent the so-called "directed acyclic graphs" based DTL technologies like Tangle or Hashgraph (Bogensperger *et al.*, 2018, p.10).

German federal government recognized that the digitalization of the nation's energy transition generally remains in its infancy. The expansion and usage of digital technology depend on developing sustainable private business models and marketable products and services. At the same time, it also requires changes in customer perceptions with regard to the usefulness of new and digitalized products. Until now, however, only smart metering can count on the establishment of a dynamic and advancing market (BMWi, 2019b, p.5). Nevertheless, digital innovations are indispensable to not only manage the growing complexity of energy generation and distribution as part of Germany's critical infrastructure (BSI, 2015, pp.107–108) but also to guarantee the functioning of Europe's wide and interconnected high-voltage power grid. Even power cuts in smaller grid areas can cause impacts and chain reactions beyond the local level and affect other European countries (Neubauer, 2020, p.724). Consequently, finding "intelligent" digital ways to manage the challenges of an ongoing decentralization, changing flow loads, a more volatile energy mix, a multiplicity of new actors, and their positioning within this increasingly dynamic energy market characterized by falling revenues is essential for the success of the Energiewende (Mika & Goudz, 2020, p.26).

Within this context and as highlighted throughout this book, the climate-related dimension must also be considered. The European Commission recognized that climate change and environmental degradation are the biggest existential threats to humankind and called for a transformation toward a more resource-efficient and competitive economy. To tackle these challenges, the EU member states committed to achieving climate neutrality by 2050. Fifty-five percent of greenhouse gas emissions relative to the 1990 level must be reduced by 2030 (European Commission, 2023b). Given that the energy sector contributes by more than 75 percent to EU GHG emissions, the roll-out of RE is key. To enhance energy security and comply with the climate-related objectives of the European Green Deal compulsory RE target of at least 32 percent for the year 2030 set by the Energy Directive 2018/2001/EU was extended to 42.5 percent by a provisional agreement between the Commission and the Council in March of 2023 (European Commission, 2023c).

8.2. SINTEG

Digital advancements such as blockchain, artificial intelligence, augmented reality, and data analytics have been named the key driving elements for the digitalization of the energy sector (BMWi, 2019b, p.15). One of the approaches to take on the challenges of the *Energiewende* represents the so-called "Smart Energy Showcase—Digital Agenda for the Energy Transition" (SINTEG) project, created to research a variety of different sectorial aspects regarding the digitalization of the *Energiewende*. According to the (then) Federal Ministry for Economic Affairs and Energy[50], SINTEG is a funding platform enabling the research and practical application of smart solutions on a case study basis within five model regions, comprising six participating states inside Germany. SINTEG supports smart energy technology testing, the development of business models, and—based on the experiences gained—recommendations for possible technical, economic, and legal changes to make the renewable-based energy transition more efficient. In other words, the goal was to create an ecosystem and an information- and communication infrastructure to meet the challenges of transforming the centralized energy system into a more decentralized one while increasing flexibility to cope with these new structural changes (BMWi, no date).

More than 300 companies, research institutions, cities and municipalities, and federal states and districts are included in the SINTEG project (WindNODE, no date: a). Analyzing the German energy transition challenges, the implementation of ICT has been detected as a necessary tool to provide a smart and secure interconnection within the increasingly decentralized energy system. Therefore, the main objective was to create a digital platform for data exchange between market participants and to increase transaction transparency for companies and consumers (WindNODE, 2020, p.62ff.). According to Hitschler and Kellermann, the German energy transition (mainly caused by a growing environmental

---

50 The Federal Ministry for Economic Affairs and Energy (BMWi) was renamed to Federal Ministry for Economic Affairs and Climate Action (BMWK) in 2021.

consciousness that resulted in changing political decision-making and legal obligations) exerts strong pressure on utility companies and municipal utilities. The energy market liberalization contributed to the entry of new actors and entities, such as energy cooperatives and wind and solar farm operators, and led to additional and volatile energy flows.

Furthermore, electricity demand is increasingly being met through own production, while energy-efficient technologies installed by industry and domestic customers have contributed to growing price competition (Hitschler & Kellermann, 2020, pp.131–132).[51] Consequently, energy-related consumer behavior and customer demand change the entire value chain, the working environment, and the culture of companies in the energy sector. Mid-size utility companies especially have to reinvent their business models and adapt them to these alterations. So, to construct and deliver a distinct *value proposition*—which can be referred to as a "holistic view of a company's bundle of products and services that are of value to the customer" (Gassmann *et al.*, 2014, p.91)—one has to detect the distinct sectors able to provide marketable value to consumers. Within the electricity markets, these "market models can be electricity, ancillary services, and/or flexibility. (…) In blockchain business models a value chain will be built on the platform connecting sellers and buyers of energy" (Talari *et al.*, 2020, p.88).

It is exactly this approach being researched by WindNODE. *Flexibility trading* has been chosen as a key focus and piece of the overall puzzle to tackle the challenges caused by the increasing volatility of (renewable) energy (WindNODE, 2020, p. 66ff.; WindNODE, 2019) and to create "a common space of trust" for the participating actors (Kirstein *et al.*, 2021, p.30). The Fraunhofer Institute for Open Communication Systems (FOKUS) has been in charge of developing intelligent electric energy solutions in the northeastern

---

51 Digital technologies are profoundly changing the traditional business model of utility companies. Thus, the decentralization of the energy sector demands applying ICT to create a digital network infrastructure between energy utility companies, grid operators, storage providers, and customers. Blockchain is one of the technologies that has gained importance within this process (Hitschler & Kellermann, 2020, p.132).

region of Germany. Its main contribution lies in the exploration and creation of ICT services, secure data, service accessibility, standardized interfaces, and open data platforms to establish an IT basis able to measure the exact inclusion of renewable resources and thereby contribute to the envisioned climate neutrality in the long run. It is within this field that blockchain solutions are being explored. The project started on December 1st, 2016, and was terminated by the end of November 2020 (Fraunhofer FOKUS, no date). The main focus of the Fraunhofer Institute's working group was thus directed toward the creation of a digital procurement aiming at the reduction of windmill output during grid congestions and the development of a flexible platform, through which the researchers expected a decreasing price effect regarding grid fees and the reduction of CO_2 emissions (WindNODE, 2018).

According to research by Knorr *et al.*, one of the main yet unresolved challenges is caused by the lack of technical options to assure the stability of the distribution grids. The digitalization, standardization, and automatization of information flows between utility companies are thereby indispensable. The latter is also necessary as no information within the existing German (and European) electricity markets regarding the exact location of energy producers as listed in the master and maturity data register exists. Only the control area is normally displayed. Consequently, a decentralized market design lacks convenient grid information for energy producers. This, however, represents a condition to allow for a more precise trade of electricity volumes, the calculation of flexibilities needed for the reduction or future elimination of bottlenecks, and the trading of so-called blind and redispatch services.

Additionally, considering production technologies allows for measuring environmental calculations (and the approximation of GHG emissions). Determining the location of standard power providers of DERs is important due to the specific weather dependency of renewable resources and the resulting difficulties in predicting their exact production level. Consequently, smart energy market solutions must include flexibility trading based on the real-time analysis of available energy resources, existing free utilization capacities, and grid, and storage capacities. Therefore, large-scale implementation of sensors is imperative (Knorr *et al.*, 2019, p.16).

The WindNODE research project at Fraunhofer Institute started by cooperating with Germany's northeastern high voltage grid operator 50Hertz (50Hertz & WindNODE, 2018). The company is in charge of operating and maintaining a grid extent of currently 10,490 kilometers within the federal states of Saxony, Thuringia, Saxony-Anhalt, Brandenburg, Mecklenburg-Western Pomerania, and the German city-states of Berlin and Hamburg. It manages voltage levels of 380 or 220 kV on long-distance transmissions between major electricity consumption nodes. In regional and highly populated areas, the grid voltage is at 110 kV. The 50Hertz customer volume encompasses around 18 million people. The company perceives itself as a frontrunner in stable RE inclusion and highlights that more than 60 percent of electricity consumption already stems from RE sources. The company's objective is to ensure a supply level of 100 percent by the year 2032. The transmission takes place at a level of 50 hertz in grids under the usage of alternating currents. However, the fluctuation resulting from stochastic energy generation and varying consumption affects the transmission frequency in power shortage and surplus times. To maintain a stable frequency and to permit a maximum oscillation of around 0.4 hertz, balancing represents one of the operator's key focuses (50Hertz online, no date).

High coordination demand exists, especially regarding energy generation and consumption data exchanges. The technical lead of the BC prototype at WindNODE highlighted that 50Hertz represents a key partner within this research due to the described large transmission network, and especially because of the highest quantity of energy transmitted among all participating entities in the research project.[52] To create the flexibility platform—the main focus of the WindNODE blockchain unit—50Hertz was therefore key in the platform's trial operation, which started in 2018. Furthermore,

52 Interview conducted with the technical lead of BC prototype on July 28th, 2020 (via an online communication platform).

Stromnetz Berlin[53], WEMAG[54], ENSO Netz[55] GmbH, and E.DIS Netz GmbH[56] partnered in the project. While these grid operators represent the biggest participants from the business sector, altogether, 70 institutions took part in the WindNODE project and encompassed producers, traders, consumers, and storage capacity developers. According to the technical lead of the BC prototype at Fraunhofer FOKUS, the exchange of national and international knowledge between researchers and actors from the private energy sector was key to developing efficient bottom-up solutions likely to serve the public good in economic and environmental terms.[57]

Within this context, creating a legal framework is particularly important to allow for a capacity calculation regarding generating and loading data for the day-ahead and intraday markets. According to Knorr *et al.*, the generation and load data provision methodology (GLDPM) represents a useful model for data transmission to comply with the system operation guideline (Knorr *et al.*, 2019, p.18). This is especially necessary considering Germany's large-scale penetration of wind and solar energy within transmission distribution grids. Additionally, changing net demand and consumption patterns (for example, toward the usage of electric vehicles) and reverse flows of energy due to the inclusion of small end users of RE can add to the beforementioned challenge of grid congestion and power voltage fluctuations (Talari, 2020, p.94). Flexible solutions are, therefore, necessary to overcome uncertainties in energy

---

[53] Stromnetz Berlin is the German capital's grid distribution operator managing voltage at 110 kv, 10 kv and 0,4 kv (Stromnetz Berlin, 2020).

[54] *WEMAG* is the grid operator for Mecklenburg-Western Pomerania, Brandenburg, and Lower Saxony with an overall operating area of approximately 8000km^2 and a voltage grid line of 15,000 km which is interconnected with 50Hertz (WindNODE, no date: b).

[55] The *ENSO Netz GmbH* is the largest energy infrastructure service provider for the eastern part of the state of Saxony, supplying around 500.000 customers with electricity and gas (WindNODE, no date: c).

[56] The *E.DIS Netz GmbH* is a public utility company who manages 35,500 km^2 of grid reaching from the Baltic Sea to the Spreewald region. Grid operations are divided into three regional areas. The company focuses on the distribution of electricity, the distribution of gas and other activities in both sections (WindNODE, no date: d).

[57] Interview conducted with the technical lead of BC prototype on July 28th, 2020 (via an online communication platform).

generation and supply. Torbahgan *et al.* refer to flexibility within energy systems "as the service to be traded in local flexibility markets in response to a market need" (Torbaghan *et al.*, 2018, p.50). In line with Minniti *et al.*, flexibility from a system perspective is defined as "(…) as the capability of the power system to adapt its production or consumption with respect to sudden changes, expected or not. At the individual level, it is often referred to as the modification of the consumption or injection pattern due to direct or indirect signals" (Minniti *et al.*, 2018, p.4). As of today, conventional generators are responsible for guaranteeing flexibility. Due to limited and insufficient ramping capacities (and the subsequent necessity to resort to spinning and frequency reserves to tackle uncertainties), flexibility products installed into the distribution and transmission grids become important in the balancing process. Batteries, electric vehicles, and demand response mechanisms are among the most preferred techniques in the system stabilization process and can be applied for tailor-made adjustments. Flexibilities in the T&D grids are offered to and regulated by the TSO to maintain the balance between generation and consumption (Talari, 2020, pp.94–95).

According to Torbaghan *et al.* (2018, p.51), there are two ahead-market scheduling mechanisms for flexibility trading locally. They are divided into *day-ahead scheduling* (AD) and *intraday scheduling* (ID). The authors argue that the:

> "day-ahead and intra-day scheduling can be utilized as long as the two corresponding auctions in the wholesale energy markets are open and accepting bids from the participants. Such a coordination between the wholesale energy market and local energy markets would serve the wholesale market participants in two ways; firstly, it would allow the wholesale market participants to maximize their profit from the wholesale energy market by inducing new production/consumption patterns in the energy program of prosumers and secondly, it would allow them to minimize their deviations from the original energy program that have been cleared in the DA and ID wholesale energy markets and the associated imbalance costs" (Torbaghan *et al.*, 2018, p.51).

For flexibility balancing to unfold its potential, Talari *et al.* (2020) argue that small flexibility resources have to be integrated. The DSO and TSO must coordinate their operations at all points to ensure that flexibility products installed inside one grid will not lead to difficulties within another. On the transmission grid basis, the main participants are traditional energy generators putting up their capacities inside reserve markets for sale. Distributed flexibility resources take on flexibility roles inside these markets and are considered secondary services inside wholesale markets. Flexibility market participants are usually large-scale aggregators offering (bigger quantities of) flexibility resources to the distribution grid. Conventional generators and transmission grid operators can also participate. Small flexibility producers submit their products to the transmission grid operator on a distribution grid level, who can then select the flexibility capacities needed at a certain price level. These flexibility products will be traded in the beforementioned day-ahead and intraday markets. Flexibility sources are thus sold for price offers in a given situation. As the blockchain element in markets for flexibility trading concentrates on the trading platform, flexibility resources have to be linked to this platform, and smart meters detect the number of flexibility resources on a real-time basis. Based on the explanations provided on specific blockchain types, *limited access private blockchains* are normally used for this trade. Blockchain technology then enables the selling of flexibility products to the T&D system operators (Talari *et al.*, 2020, pp.95–96).

8.3. WindNODE Flexibility Platform

The particularities of the German interregional transmission system and the options to install blockchain-based solutions shall now be considered. This is important as the expansion of primary RE generation grew at a much faster pace than initially envisioned. Their share in cross-final energy consumption was 20.4 percent in 2022, while – driven by the growing electrification demand – the share of renewables in electricity consumption rose to 46.2 percent (German Federal Environmental Agency, 2023, p.6). However, a stronger contribution by renewables has been limited due to an insufficient grid expansion. Thus, in the second quarter of 2022 alone, absolute reductions of RE-generated electricity amounted to 2.134 GWh. This represents an increase of around 40 percent relative to the same timeframe in 2021. However, the compensation rights amounted to "only" 56 Mio. EUR, which was significantly lower than in the second quarter of 2021 (BNetzA, 2022, pp.3–4).[58] With 36.9 (or 788 GWh) and 50.1 percent (or 1.069 GWh), respectively, of on-and offshore wind energy, this major RE source shows the highest curtailment level (BNetzA, 2022, p.21). These data highlight the necessity to implement ICT technologies for data exchanges between grid operators, plant operators, and consumers in the shape of smart markets to balance demand and supply more efficiently.

Electricity trading in Germany and Europe-wide is based on a "fictitious" unconstrained functioning of grid operations. Long-term trading is conducted on the market for futures, while short-term deals are concluded via the EPEX Spot market (day-ahead and intraday trading). The stochastic generation of RE sources is increasingly causing difficulties in correctly predicting generation and balancing. Consumer behavior also became less forecastable. Thus, the control energy market represents an economic tool to manage future imbalances within defined control regions and is subdivided into primary, secondary, and minute reserves. The TSO displays the necessary predicted amount of additional flexibility on a platform on a daily or weekly basis, and plant operators can offer these volumes with a defined energy price on a pay-as-bid pricing model. If imbalances occur, TSOs can resort to these resources based on the merit-order-curve principle (Knorr *et al.*, 2019, pp.67–68; BNetzA, 2019, pp.25–26).

However, given the growing fluctuations of demand and supply and constant grid interventions (coming along with a cost-intensive redispatch and feed-in management), a new flexible trading design focusing on short-term trading with the option to include a wide array of energy producers and resources needs to be created. Furthermore, this process has to consider sector coupling and load management based on incentive-driven market mechanisms to

58 This is due to legal changes regarding the feed-in management. While the latter has formerly been part of the renewable energy act (EEG), the redispatch regulations (or so called "Redispatch 2.0") have been redefined and entered into the Energy Industry Act. As a result, operators of RES are only paid the so-called "market premium" (BNetzA, 2022, pp.3–4).

generate competition between flexibility options. However, the current merit-order curve model and grid feed-in obligations of RE sources lead to investment considerations primarily based on low operating cost, yet they disregard the necessities of local grid situations (Knorr, 2019, pp.70–71). According to a study by Ecofys und Fraunhofer IWES (2017) commissioned by AGORA Energiewende on smart market designs in the German distribution networks, two smart market models, which are also referred to as "traffic light approaches" have to be taken into account. The latter combines specific physical networks based on technical and economic considerations in the yellow phase (BNetzA, 2017, pp.14–16). These smart market models can be subdivided into two categories: 1. *Quota* and 2. *Flexibility* models.

The first category focuses on restricting energy producers and their clients based on congestion predictions. The quotas can be traded between market participants, creating a market price based on price preferences within these smart markets. The flexibility category refers to a mechanism in which operators act as sole demanders within the regional energy market and reduce and increase power levels until a balance is established. There are currently two quotas (regulated pricing and voluntary quota) and four flexibility models (cascade, regional reserve MarketPlus, regional IntradayPlus, and new flexibility platform). According to the cascade model, the connected actors give information about their capacities to provide flexibility at any moment. As shown before, the distribution operators or independent entities can run the platform. The platform can activate connected power suppliers based on supply and demand or congestion predictions. Within the regional reserve MarketPlus, the increased reserve control products are based on geographic considerations, thereby paying attention to distinct local producers and grid conditions. The DSO is running this flexibility platform. The regional IntradayPlus market is similar. However, it has an added intraday trading option, while the DSO or TSO act as demanders. The new flexibility platform provides access to all system operators for congestion management and grid stability purposes but remains unrelated to normal market operations (Ecofys & Fraunhofer IWES, 2017, pp.22–27).

According to the technical lead of BC prototype of the blockchain section at WindNODE, their platform was developed based upon the *new flexibility platform model*. Ethereum, in its most recent version, had been chosen. Interestingly, although PoW is being used, the researchers use a private, permissioned blockchain on the Ethereum platform.[59] To enable the trading process, smart contracts have been written in Solidity based on PoW with an adapted level of difficulty to mine the block. Unfortunately, the domain-specific language of Solidity, although the most used to program smart contracts, is the least secure of all currently existing languages and prone to vulnerabilities (Parizi *et al.*, 2018, p.90). This underlines the limitations of secure blockchain usage. A process of tokenization was not further pursued. Generally, however, it is possible to distribute tokens after this transaction, which can be coupled with Fiat currencies.[60] The flexibility trading could, therefore, also be bound to the Euro. The technical lead of the BC prototype regards the latter as a more complex yet possibly worthwhile endeavor, as digital currency fluctuations could be ruled out. These extensions, however, depend on the progress and acceptance of blockchain applications on behalf of producers and customers. As a result, the Fraunhofer FOKUS blockchain unit currently focuses on the feasibility of detecting supply and demand, the general analysis of behavioral patterns, and the measurement and availability of storage capacities to reduce congestion.[61]

By doing so, the researchers applied sensors enabling big data and artificial intelligence (AI) to take on the non-linear and uncertainty-related difficulties within the traditional German energy system. Their goal is to establish what Xu *et al.* call an overall "platform thinking to develop a holistic view on the AI energy platform" (Xu *et al.*, 2019, p.16).

---

[59] Online communication conducted with the technical lead of BC prototype on September 29th, 2020.

[60] Online communication conducted with the technical lead of BC prototype on September 29th, 2020.

[61] Online communication conducted with the technical lead of BC prototype on September 29th, 2020.

At the time of implementation, the blockchain platform ran (as outlined) on the more energy-intensive PoW consensus on the Ethereum platform. The resulting high CO_2 emissions due to mining activities at the time led the researchers to consider future switches to the PoS consensus to avoid emissions they intended to save through the platforms' creation and implementation in the first place.[62] With Ethereum switching to PoS in 2022, reality has caught up with these considerations.

Consequently, while the test phase, which started in 2018, showed positive results in reducing bottlenecks, they are not yet cost-efficient enough to justify the investments into the decentralized and digitalized infrastructure necessary to implement the blockchain-based flexibility platforms.[63] More participants must join the platform and be willing to share their energy data to be more efficient. According to the technical lead of BC prototype, this currently represents the biggest challenge because many players within the market do not want to share these data for privacy protection and economic reasons.[64]

8.4. Conclusion

This chapter has shown that the German energy transition comes with key challenges to balance the supply and demand of RE within regional and interregional transmission grids. The volatile generation characteristics of RE and their weather dependency continue

62 The PoW consensus is considered to be the most energy-intensive. As suggested by de Vries et al., Bitcoin mining may be responsible for 65.4 $MtCO_2$ annually, comparable to the CO_2 emissions of a country like Greece (de Vries *et al.*, 2022). However, arguments made with regards to the negative environmental impact of bitcoin mining are challenged by an emerging body of literature, arguing that PoW-mining can actually make a substantial contribution to the greening of the Bitcoin network by providing incentives to invest in RES (Ibañez & Freier, 2023; Cross & Baily, 2021).

63 Interview conducted with the technical lead of BC prototype on July 28th, 2020 (via an online communication platform).

64 Interview conducted with the technical lead of BC prototype on July 28th, 2020 (via an online communication platform).

to lead to large-scale electricity losses. However, these stochastic generation patterns of renewable resources affect the national and interconnected European grids. Thus, to help balance demand and supply, researchers from the Fraunhofer Institute have developed a flexible platform on a blockchain basis, allowing for near real-time responses to avoid bottlenecks and stabilize the grid. A first trial run started in 2018 and showed positive results in reducing energy shortages. Functioning on market logic, the creation of the blockchain-based smart market showed the first positive indications that a balancing effect is occurring.

While creating this regional trading platform aimed to balance energy flows and sufficiently incorporate renewables in the grid, it can also be perceived as a bottom-up approach contributing to reducing GHG emissions in the energy sector. The latter, however, has not yet made significant progress. This can also be attributed to the relatively small number of private and public entities participating in the research project. So, the Fraunhofer research showed that new technologies, such as blockchain, must first gain acceptance within society. This is so as the successful implementation of blockchain-based flexibility trading calls for the participation of most private and public actors. At this point, however, the importance of privacy protection considerations is seemingly dominating Germany over the global common good of making renewable integration more efficient. Therefore, political decision-making and the design of these technical innovations need to focus on meeting these very concerns to allow for more actors in blockchain-based digital trading platforms.

Finally, such a perception is closely connected to developing an appropriate legal framework. A spillover of technological innovations onto the development of efficient legal instruments to promote and apply blockchain-based innovations to the energy sector continues to be slow due to varying economic and political interests. Consequently, they are expected to be fully implemented on a European scale only if blockchain technology in the energy sector proves to be an efficient and economically viable tool.

From a theoretical standpoint, policy learning, policy diffusion, and technology transfer and a lack thereof can best describe the developments surrounding the creation and application of new technologies to the German energy sector. Existing international knowledge and the exchange of ideas, especially on technological advancements, represent the basis for international research cooperation in the contemporary world. Furthermore, Fraunhofer's cooperation partners are large economic players concentrating on international expertise to carry out their operations and learn from other countries' experiences. The main challenge is, to a lesser extent, the creation of blockchain-based solutions, but their actual application due to a lack of normative and legal adaptations. So, while technological advancements exist (at least partially) to overcome the challenges resulting from the volatile renewable generation and to enhance the effective use of flexibilities to increase the balancing, it is the lack of trust on behalf of energy producers and consumers that the provided information shall not be misused. So, changing this culturally anchored distrust with regard to possible abuse of information has to be tackled to provide for a reduction of CO_2 emissions stemming from the energy sector.

Consequently, while international negotiations promoted by regime theory and the institutional strands of green theory to solve international challenges enhance policy learning and technology transfer, this chapter shows that specific local conditions keep technological innovations from being applied. Unique individual rights, like safeguarding personal data, come into conflict with the imperative of data collection, essential for advancing a sustainable low-carbon energy industry and significant environmental progress. These two facets aim to enhance the greater common good. Inasmuch, different social actors must be aware of this clash and discuss policy approaches, allowing compromises to make both ends work without jeopardizing fundamental rights or policy objectives. The spread of blockchain, its further sophistication, tamper-proof, and encrypted nature might as well be at the heart of solving these challenges. Therefore, awareness of how this technology works and where it can be improved is of key interest to society.

9. Main Conclusion

This book sheds light on the complexity of breaking through the vicious cycle characterizing the global climate change—fossil fuel energy nexus and the difficulties of effective solutions to properly implement RE resources as a viable alternative to today's continuous petroleum and gas dependency. This is even more essential given that countries conducting comprehensive policy changes toward transforming their national energy systems also struggle to reduce their GHG missions.

To counter this urgent challenge, the digitalization of the global energy sector has increasingly found entry into the academic debate and public policy discussion to find appropriate smart solutions to the problem of volatile energy generation, especially from resources such as wind and solar. The latter not only represent the key resources to transform energy matrices in countries with ambitious RE and decarbonization objectives to comply with international climate contracts; digital innovations are also regarded as necessary to take on the distinct characteristics of most RE resources. Due to their distinct weather dependency, wind farms and solar panels are not always producing energy at times when needed for consumption. Thus, a sufficient backup mechanism to store energy generated from renewable resources does not exist. These difficulties are enhanced in some countries by the additional closure of nuclear power plants. Although an economically stable and low-carbon energy resource, this is due to growing political pressure arising from nuclear disasters and reoccurring plutonium leakages at different global nuclear energy sites (Sinn, 2017; Sinn, 2012).

Energy security, understood by Yergin (2006) as the affordable, reliable, and unconstrained access to energy as the macroeconomic basis for a country's economic functioning and growth, must be guaranteed at all times. Supply chain interruptions and energy price changes can lead to economic disadvantages and affect a nation's competitive positioning within the global economy. Inasmuch, energy access entails an economic but subsequently also social dimension. High energy costs must be refinanced and might lead to increased taxes, fees, and contributions while representing an additional social burden often difficult to bear for large parts of society. As a result, energy poverty became a phenomenon not only characterizing developing countries.

Focusing on digital means to combat GHG emissions and maintain a country's competitive economic edge represent two key challenges within contemporary energy governance. Among one of the most discussed digital innovations is blockchain. While having shown signs of disrupting the financial industry due to new financial transaction models, this DLT represents at its core a shared and distributed database that may contain digital transactions, data records, and executables aggregated into larger formations (blocks) which are cryptographically linked to previous blocks forming a chain of records (Andoni *et al.*, 2019, p.145). One key idea behind blockchain's application to the energy sector is to allow for energy trading while measuring the demand and supply of energy sources to prevent bottlenecks. Based on the data obtained, the applied ICT infrastructure shall permit information exchange regarding further energy sources and available storage capacities.

Derived from these observations and departing from an international relations theory perspective with regard to the usage of RE as a means aiming at the reduction of GHG emissions, this book looked specifically at global bottom-up strategies of blockchain applications to the RE sector and analyzed them for their economic feasibility. By doing so, this research was based upon a multi-academic approach and called for the inclusion of an international relations and legal analysis to answer the underlying hypothesis.

The latter hypothesis claimed that blockchain technology could be a viable option to both enhance the efficiency and balancing between RE supply and demand and subsequently reduce GHG emissions if the three components of 1. technological advancements, 2. an adequate global normative framework, and 3. a general level of trust on behalf of key market actors to promote blockchain energy on a global scale, exists.

So, to approach the overall theme of this book theoretically and to prove the hypothesis, four key theories or theoretical concepts have been selected. First of all, Global Governance theories, as outlined at the beginning of the theoretical chapter, highlighted the fact that in an increasingly interconnected world with diverse threat potential only solvable through concerted action and a general shift in the focus of research and policy shall allow for the creation of a more just global society and the mitigation of conflicting agendas. As argued by Mayntz (2002), in an increasingly interconnected world, these diverse conflict potentials overlap while being likely to multiply and manifest themselves at critical junctures. The latter are indicators that global climate change and historically grown and diverse political, economic, and social configurations are colliding.

As mentioned in this book's introduction, environmental protection and sufficient access to sustainable energy sources represent basic rights and should be included in a more comprehensive human-security approach as proposed by the United Nations. The question raised by Oran Young (2005), therefore, as to why there is no unified theory of global environmental governance can thereby be answered. Global climate change represents a highly multifaceted topic. It involves political, economic, social, and environmental dimensions and diverging ideological perceptions on tackling them best. And it is the incompatibility of thinking exclusively nationally while trying to solve global problems that call for a rethinking of global cooperation and a strong international consensus regarding new developments, especially in the energy sector.

Since its commercial exploitation and usage from the 19th century onward, petroleum has been the basis for the functioning of the global economy, which led to an impressive push in human development and unprecedented effects on innovation. However, as the energy-climate change nexus, as analyzed in Chapter Four has shown, this model has become unsustainable at the current pace of industrial development and global competition. Consequently, a new consensus must be found to develop effective contributions among state, private, and societal actors on a global level.

The continuously growing GHG emissions serve as a strong indicator that such consensus does not yet exist. So, to find a suitable solution for the energy sector as the prime contributor to global GHG emissions, the community must adopt appropriate regional, federal, and international adjustment measures while considering the global common good.

Liberal institutionalism as a so-called 'problem-solving theory' shows that international cooperation and creating international regimes, such as climate change Agreements, represent opportunities to coordinate state policies and benefit from collective action. However, it also highlights the challenges of policy coordination due to diverging interests and fears of relative political and economic losses of states. In contrast to classic security-related issues, the threat of climate change appears yet too abstract by some states to agree on sovereignty transfer and the creation of efficient international coordination. Effective concessions to maintain global warming well below the intended two-degree Celsius level proposed by the Paris Agreement are therefore only visible to a limited extent. Policy learning and diffusion approaches have emerged as key strategies to discuss and exchange ideas on governance concepts and technology transfer. As Chapter Five highlights, international climate contracts such as the Kyoto Protocol and the Paris Accord have recognized the importance of transferring low-carbon or climate-friendly technologies. Therefore, technological innovations like those in the RE sector can be directly derived from climate texts. The Paris Agreement particularly calls for more flexibility with regard to the consideration of specific local contexts, which demand the recognition of distinct local geographic, climate, political, economic, and social conditions.

Bottom-up strategies such as using "blockchain energy" for microgrids and long-distant electricity transmission management can be considered a possible solution if the overall price efficiency (compared to other resources) and the climate-friendliness of these innovations can be proven. Such a take on the issue aligns with the liberal-institutionalist position within both regime theory and the anthropocentric approaches of green theory, according to which activating human engagement is based upon incentive structures.

Policy and technological transfer, however, demand trust among key economic and governmental actors to exchange information regarding state-of-the-art technologies mainly developed in industrialized countries in the Northern Hemisphere. However, given the competitive logic of the global economy and existing challenges to guarantee intellectual property rights in many countries, private actors, in particular, remain very reluctant to share their knowledge fully. So, these actors' more profound engagements are still obscured by uncertainties of possible free riding and its relative economic consequences on countries developing innovations such as in the field of blockchain.

This becomes even more apparent considering the cost of conducting global changes in the energy sector. The IEA and IRENA estimate necessary annual investments in RE between 2018 and 2030 to be at around $55 billion to increase energy access, $700 million to increase available RE resources, and $600 million to improve EE to meet the sustainable energy goals within the area of green energy (World Bank, 2019, p.1). This burden will primarily have to be shared by industrialized countries of the Global North.

As this research has shown in Chapter Six, blockchain is an investment-intensive technology, which—to be globally spread and applied to the distinct local contexts—needs to come along with the development of further smart technologies such as DER, ICT, smart contracts, sensors, etc.

One of the main advantages of blockchain's DLT is that it allows for the emergence of decentralized business models by removing the need for intermediaries or central management. Such an implication presents blockchain as an attractive model from an economic point of view.

As this research has shown, blockchain applications in the energy sector have various advantages when applied to microgrids. In these cases, local RE exchanges come with high approval rates among participants and contribute to a strengthened sense of community unity. While this personal and experience-based perspective is important, it can—interestingly—be enhanced by rational and incentive-based economic interests in sustainable business model developments. Therefore, this research highlights that economic and interest-driven ideas on business maximization in the green energy and climate protection domain and personal, well-being-oriented considerations are not mutually exclusive. They have the potential to strengthen and reinforce each other.

While prosumers within the BMG context have become engaged in contributing to the community in social and environmental terms, additional options to earn money through selling privately operated and rooftop-based solar energy within the microgrid have increasingly become an economic factor. So, the project design by LO3 Energy in coordination with Siemens has shown to be a self-sustaining strategy. The basic blockchain trading mechanism needs to be improved to substantially contribute to real-time energy exchange and more effective long-term GHG emission reductions. So, the idea of blockchain to generate trust, transparency, and democratization on a technical level displayed spillover effects on personal relations and reinforced the overall sense of community. Such an actors-oriented approach applies to the theoretical theme chosen within this book and underscores the growing importance of bottom-up initiatives to both meet the challenges of energy security and the necessity to reduce GHG emissions. Energy transitions involving consumers and prosumers as small-scale participants on a community and real-time basis are viable approaches to tackle the distribution challenges and better react to distinct local production and consumption patterns.

The WindNODE project, on the other hand, cannot count on such support. Instead, it has been engaged with distrust and is considered mainly a political-economic tool. For it to improve, large-scale investments into the entire blockchain infrastructure by public and private actors, its trading capacity, and a reduction of the "scalability trilemma" have to be developed. Potentially, it is an interesting idea to balance supply and demand and contribute to future GHG reductions. At the current state of development, however, this analysis has shown that the blockchain section within WindNODE does not live up to the intended objectives regarding its climate dimensions.

Deriving from this book's research results, the outlined hypothesis—according to which blockchain technology can be a viable option to both enhance the efficiency and balancing between RE supply and demand and subsequently reduce GHG emissions, if the three components of 1. technological advancements, 2. an adequate global normative framework, and 3. a general level of trust on behalf of key market actors to promote blockchain energy on a global scale, exists—can be proven.

10. Bibliography

ABI Research (2020). "Blockchain Markets Take a US$2.8 Billion Hit." URL: https://www.abiresearch.com/press/blockchain-markets-take-us28-billion-hit/ (accessed 16.05.2023).

Adams, Samuel; & Nsiah, Christian (2019). "Reducing carbon dioxide emissions: Does renewable energy matter?" Science of the Total Environment. Vol. 693. DOI: 10.1016/j.scitotenv.2019.07.094.

Adefarati, Temitope; & Bansal, Ramesh (2019). "Energizing Renewable Energy Systems and Distribution Generation." In: Taşcıkaraoğlu, Akin; Erdinç, Ozan (Editors). Pathways to a Smarter Power System. Academic Press/Elsevier: London. pp. 29–65. DOI: 10.1016/B978-0-08-102592-5.00002-8.

Adger, W. Neil; Pulhin, Juan M.; Barnett, Jon; Dabelko, Geoffrey D.; Hovelsrud, Grete K.; Levy, Marc; Spring, Úrsula Oswald; & Vogel, Coleen H. (2014). "Human security." In: Climate Change 2014: Impacts, Adaptation, and Vulnerability. Part A: Global and Sectoral Aspects. Contribution of Working Group II to the Fifth Assessment Report of the Intergovernmental Panel on Climate Change. In: Field, Chris B.; Barros Vicente Barros; Dokken, David Jon; Mach, Katharine J.; Mastrandrea, Michael D.; Bilir, Taha Eren; Chatterjee, Monalisa; Ebi, Kristie L.; Estrada, Yacob Olani; Genova, Rafael Coloma; Girma, Baris; Kissel, Ellen S.; Levy, Ana N.; MacCracken, Steve; Mastrandrea, Pamela R.; White, Leo L. (eds.). Cambridge University Press, Cambridge, United Kingdom and New York, NY, USA, pp. 755–791. URL: https://www.ipcc.ch/site/assets/uploads/2018/02/WGIIAR5-Chap12_FINAL.pdf.

AGORA Energiewende; & Guidehouse (2021). "Making renewable hydrogen cost-competitive: Policy instruments for supporting green H_2." Version: 1.2, August 2021. Berlin. pp. 1–96.

AGORA Energiewende, & Enervis (2021). "Phasing out coal in the EU's power system by 2030. A policy action plan." Impuls. Version: 1.0, October 2021. Berlin.

Agnew, John (2005). "Sovereignty Regimes: Territoriality and State Authority in Contemporary World Politics." Annals of the Association of American Geographers, Vol. 95(2), pp. 437–461. DOI: 10.1111/j.1467-8306.2005.00468.x.

Akhmat, Ghulam; Zaman, Khalid; Shukui, Tan; & Sajjad, Faiza (2014). "Does energy consumption contribute to climate change? Evidence from major regions of the world." Renewable and Sustainable Energy Reviews, Vol. 36, pp. 123–134. DOI: 10.1016/j.rser.2014.04.044.

Alfredsson, Eva C. (2014). '"Green' consumption–no solution for climate change." Energy. Vol. 29 (4). pp. 513–524. DOI: 10.1016/j.energy.2003.10.013.

Alter, Karen J.; & Raustiala, Kal (2018). "The Rise of International Regime Complexity." Annual Review of Law and Social Science. Vol. 14, pp. 329–349. DOI: 10.1146/annurev-lawsocsci-101317-030830.

Ali, Mumtaz; & Seraj, Mehdi (2022). "Nexus between energy consumption and carbon dioxide emission: evidence from 10 highest fossil fuel and 10 highest renewable energy-using economies." Environmental Science and Pollution Research. Vol. 29 (58), pp.87901–87922. DOI: 10.1007/s11356-022-21900-9.

Anderson, Terry L.; & Leal, Donald R. (2001). "Free Market Environmentalism." New York: Palgrave Macmillan.

Andoni, Merlinda; Robu, Valentin; Flynn, David; Abram, Simone; Geach, Dale; Jenkins, David; McCallum, Peter; & Peacock, Andrew (2019). "Blockchain technology in the energy sector: A systematic review of challenges and opportunities." Renewable and Sustainable Energy Reviews. Vol. 100, pp. 143–174. DOI: 10.1016/j.rser.2018.10.014.

Arduin, Pierre-Emmanuel (2021). "A cognitive approach to the decision to trust or distrust phishing emails." International Transactions in Operational Research. pp. 1–36. DOI: 10.1111/itor.12963.

Banda, Maria L. (2018). "The bottom-up alternative: The mitigation potential of private climate governance after the Paris Agreement." Harvard Environmental Law Review. Vol. 42., pp. 325–389.

Bard, Jochen; Gerhardt, Norman; Selzam, Patrick; Beil, Michael; Wiemer, Martin; & Buddensiek, Maike (2022). "The limitations of hydrogen blending in the European gas grid. A study on the use, limitations, and cost of hydrogen blending in the European gas grid at the transport and distribution level." Fraunhofer Institute for Energy Economics and Energy System Technology (IEE). pp. 1–50. DOI:10.13140/RG.2.2.30093.41448.

Bayer, Patrick; & Aklin, Michaël (2020). "The European Union Emissions Trading System reduced CO2 emissions despite low prices." Proceedings of the National Academy of Science of the United States of America. Vol. 117 (16), pp. 8804–8812. DOI: 10.1073/pnas.1918128117.

Bennett, David; Liu, Xiaming; Parker, David; Steward, Fred; & Vaidya, Kirit (2001). "Technology Transfer to China: A Study of Strategy in 20 EU Industrial Companies." International Journal of Technology Management. 21(1-2), pp. 151–182. DOI: 10.1504/IJTM.2001.002899.

Beunen, Raoul; Van Assche, Kristof; & Gruezmacher, Monica (2022). "Evolutionary Perspectives on Environmental Governance: Strategy and the Co-Construction of Governance, Community, and Environment." Sustainability. Vol. 14 (16), 9912. pp. 1–18. DOI: 10.3390/su14169912.

Beveridge, Ross; & Kern, Kristine (2013). "The 'Energiewende' in Germany: Background Developments and Future Challenges." Renewable Energy Law and Policy Review. Vol. 4(1). Special Issue. Grid and Energy Infrastructure. pp. 3–12.

Błaszczyk, Michał; Popović, Milan; Zajdel, Karolina; & Zajdel, Radosław (2022). "The Impact of the COVID-19 Pandemic on the Organisation of Remote Work in IT Companies." Sustainability. Vol. 14(20), 13373, pp. 1–14. DOI: 10.3390/su142013373.

BMWi (no date). "Funding programme Smart Energy Showcases—Digital Agenda for the Energy Transition" (SINTEG). URL: https://www.bmwk.de/Redaktion/EN/Artikel/Energy/sinteg-funding-programme.html (accessed 4.06.2023).

BMWi (2019a). "Kommission Wachstum, Strukturwandel und Beschäftigung.". Abschlussbericht. Berlin.

BMWi (2019b). „Gutachten. Digitalisierung der Energiewende. Topthema 1: Verbraucher, Digitalisierung und Geschäftsmodelle". Berlin, pp. 1–66.

BNetzA (2022). "Quartals-Bericht. Netzengpassmanagement Zweites Quartal 2022." pp. 1–28. URL: https://www.bundesnetzagentur.de/SharedDocs/Downloads/DE/Sachgebiete/Energie/Unternehmen_Institutionen/Versorgungssicherheit/Engpassmanagement/QuartalszahlenQ2in2022.pdf?__blob=publicationFile&v=3 (accessed 9.05.2023).

BNetzA—Federal Network Agency (2019). "Begleitdokument zur Methode der ÜNB der Kapazitätsberechnungsregion Hansa für ein marktbasiertes Verfahren zur Zuweisung grenzüberschreitender Übertragungskapazität für den Austausch von Regelleistung oder die Reserventeilung gemäß Artikel 41 der Verordnung (EU) 2017/2195 der Kommission vom 23. November 2017 zur Festlegung einer Leitlinie über den Systemausgleich im Elektrizitätsversorgungssystem (18.12.2019)." pp. 1–46. URL: https://www.bundesnetzagentur.de/DE/Service-Funktionen/Beschlusskammern/1_GZ/BK6-GZ/2019/BK6-19-567/BK6-19-567_Begleitdokument_2019_12_18.pdf?__blob=publicationFile&v=1 (accessed 18.9.2020).

BNetzA—Federal Network Agency (2017). "Flexibility in the electricity system. Status quo, obstacles and approaches for a better use of flexibility." Discussion paper Translation of German version published in April 2017. pp. 1–55. URL: https://www.bundesnetzagentur.de/SharedDocs/Downloads/EN/Areas/ElectricityGas/ FlexibilityPaper_EN.pdf?blob=publicationFile&v=2 (accessed 18.09.2020).

Bogensperger, Alexander; Zeiselmair, Andreas; & Hinterstocker, Michael (2018). "Die Blockchain Technologie. Chance zur Transformation der Energieversorgung? Berichtsteil Technologiebeschreibung." Forschungsstelle für Energiewirtschaft.pp. 1–90.

Böhringer, Christoph; & Rutherford, Thomas F. (2007). "Combining Top-Down and Bottom-up in Energy Policy Analysis: A Decomposition Approach." Discussion Paper No. 06-007. URL: https://www.mpsge.org/qpdecomp.pdf. (accessed 20.04.2020).

Bookchin, Murray (2006). "Social Ecology and Communalism." Oakland: AK Press.

Bookchin, Murray (1996). "The Philosophy and Social Ecology. Essays on Dialectical Naturalism." Second Edition. Black Rose Books. URL: https://theanarchistlibrary.org/library/murray-bookchin-the-philosophy-of-social-ecology. (accessed 20.05.2022).

Botene, Pedro Henrique Ribeiro; De Azevedo, Anibal Tavares; & De Arruda Ignácio (2021). "Blockchain as an enabling technology in the COVID-19 pandemic: a systematic review." Health & Technology 11, pp. 1369–1382. DOI: 10.1007/s12553-021-00593-z.

BP (2022). "BP Statistical Review of World Energy 2022." 71st edition. URL: https://www.bp.com/content/dam/bp/business-sites/en/global/corporate/pdfs/energy-economics/statistical-review/bp-stats-review-2022-full-report.pdf (accessed 06.04.2023).

BP (2019). "BP Statistical Review of World Energy 2019." 68th Edition. URL: https://www.bp.com/content/dam/bp/business-sites/en/global/corporate/pdfs/energy-economics/statistical-review/bp-stats-review-2019-full-report.pdf (accessed 25.05.2020).

Bradshaw, Michael (2010): "Global energy dilemmas: a geographical perspective." The Geographical Journal, Vol. 176(4), pp. 275–290. DOI:10.1111/j.1475-4959.2010.00375.x.

Brunnée, Jutta (2003). "The Kyoto Protocol: A Testing Ground for Compliance Theories?" Heidelberg Journal of International Law. Vol. 63 (2), pp. 255–280.

Brunner, Ronald (2001). "Science and the climate change regime." Policy Science. Vol. 34 (1), pp. 1–33. DOI:10.1023/A:1010393101905.

BSI (Federal Office for Information Security) (2015). "KRITIS–Sektorstudie. Energie." Öffentliche Version – Revisionsstand 5. February 2015. Berlin.

Bueno, Maria del Pilar (2019). "Identity-Based Cooperation in the Multilateral Negotiations on Climate Change." In: Lorenzo, Cristian (editor). Latin America in Times of Global Environmental Change. Cham: Springer. pp. 57–73. DOI: 10.1007/978-3-030-24254-1_5.

Busch, Per-Olof; & Jörgens, Helge (2012). "Governance by diffusion: exploring a new mechanism of international policy coordination." In: Langhelle, Oluf; Meadowcroft, James; Ruud, Audun (editors). Democracy, Governance and Sustainable Development: Moving Beyond the Impasse: Edward Elgar Publishing, pp. 221–248. DOI: 10.4337/9781849807579.00019.

Business Insider (2020). "CO2 European Emission Allowances Commodity" (21.09.2020) URL: https://markets.businessinsider.com/commodities/co2-european-emission-allowances (accessed 21.09.2020).

Cadman, Timothy (2013). "Introduction: Global Governance and Climate Change." In: (idem). Global climate change policy: towards institutional legitimacy. New York: Palgrave Macmillan pp.1−−16. DOI: 10.1057/9781137006127_1.

Carlarne, Cinnamon P. (2012). "Rethinking A Failing Framework: Adaptation and Institutional Rebirth for the Global Climate Change Regime." Georgetown International Law Review. Vol. 25 (1). 1–50.

Casey, Michael J.; & Vigna, Paul (2018). "The Truth Machine. The Blockchain and the Future of Everything." Macmillan: New York.

Casola, Laura; & Freier, Alexander (2018). "El nexo entre cambio climático y energía renovable en el Mercosur. Un análisis comperativo de las legislaciones de Argentina y Brasil." Revista Derecho Del Estado. Vol. 40, 153–179. DOI: 10.18601/01229893.n40.07.

Chalendar, Jaques A.; & Benson, Sally M. (2019). "Why 100% Renewable Energy Is Not Enough." Joules. Vol. 3 (6). pp. 1389–1393. DOI: 10.1016/j.joule.2019.05.002.

Conway, Declan; Nicholls, Robert J.; Brown, Sally; Tebboth, Mark G.L.; Adger, William Neil; Ahmad, Bashir; Biemans, Hester; Crick, Florence; Lutz, Arthur F.; De Campos, Ricardo Safra; Said, Mohammed; Singh, Chandni; Zaroug, Modathir Abdalla Hassan; Ludi, Eva; New, Mark; & Wester, Philippus (2019). "The need for bottom-up assessments of climate risks and adaptation in climate-sensitive regions." Nature Climate Change. Vol. 9 (7). pp. 503–511. DOI: 10.1038/s41558-019-0502-0.

Crosby, Michael; Nchiappan; Pattanayak, Pradhan; Verma, Sanjeev; & Kalyanaraman, Vignesh (2015). "Blockchain Technology beyond Bitcoin." Sutardja Center for Entrepreneurship & Technology Technical Report. Berkeley University. pp. 1–35.

Cross, Troy; & Bailey, Andrew M. (2021). "Greening Bitcoin with Incentive Offsets." Whitepaper. URL: https://docs.google.com/document/d/1N2N-5jY00cmteoY_puWI9oosM1foa4EQqsO1FFfIFR4/edit (accessed 9.05.2023).

Cunha Álvaro; Martins, Jorge; Rodrigues, Nuno; & Brito, Francisco (2014). "Vanadium redox flow batteries: a technology review." International Journal of Energy Research, Vol. 39(7), pp. 889–918. DOI: 10.1002/er.3260.

Cutler, Claire (2003). "Private Power and Global Authority. Transnational Merchant Law in the Global Political Economy." Cambridge Studies in International Relations. Cambridge; Cambridge University Press. DOI: 10.1017/CBO9780511550300.

Cutler, Claire (2004). "Private international regimes and interfirm cooperation." In: Hall, Rodney Bruce & Biersteker, Thomas J. (Editors). The Emergence of Private Authority in Global Governance. Cambridge: Cambridge University Press. pp. 23–40. DOI: 10.1017/CBO978051149 1238.003.

Dalby, Simon (2013). "Climate Change as an issue of human security." In: Redclift, Michael R.; & Grasso, Marco (editors). Handbook on Climate Change and Human Security. Northampton: Edward Elgar Publishing. pp. 21–40. DOI:10.4337/9780857939111.00009.

De Filippi, Primavera; Morshed, Mannan; & Reijers, Wessel (2020). "Blockchain as a confidence machine: The problem of trust & challenges of governance." Technology in Society. Vol. 62. pp. 1–14. DOI: 10.1016/j .techsoc.2020.101284.

De Vries, Alex; Gallersdörfer, Ulrich; Klaaßen; & Stoll, Christian (2022). "Revisiting Bitcoin's Carbon Footprint." Joule. Vol. 6 (13), pp. 498–502. DOI: 10.1016/j.joule.2022.02.005.

Derbali, Abdelkader; Jamel, Lamia; Mani, Yosra; & Al Harbi, Raied (2019). "How will Blockchain change corporate governance." International Journal of Business and Risk Management. Vol. 2 (1), pp. 16–18. DOI: 10.12691/ijbrm-2-1-3.

Demeritt, David (2006). "Science studies, climate change and the prospects for constructivist critique." Economy and Society. Vol. 35(3), pp. 453–479. DOI: 10.1080/03085140600845024.

Dirix, Jo; Peeters, Wouter; Eyckmans, Johan; Jones, Peter Tom; & Sterckx, Sigrid (2013). "Strengthening bottom-up and top-down climate governance." Climate Policy. Vol. 13 (3), pp. 363–383. DOI: 10.1080/1 4693062.2013.752664.

Di Silvestre, Maria Luisa; Gallo, Pierluigi.; Ippolito, Mariano Guiseppe; Riva Sanseverino, Elenora; Sciumè, Guiseppe; & Zizzo, Gaetano (2018). "An Energy Blockchain, a use case on Tendermint." Department of Energy. The University of Palermo. IEEE International Conference on Environment and Electrical Engineering and 2018 IEEE Industrial and Commercial Power Systems Europe. pp. 1–5. DOI: 10.1109/EEEIC.2018.8493919.

Dolowitz, David P; & Marsh, David (2000). "Learning from Abroad. The Role of Policy Transfer in Contemporary Policy-Making." Governance: An International Journal of Policy and Administration. Vol. 13 (1). pp. 5–24. DOI: 10.1111/0952-1895.00121.

Dobson, Andrew (2007). "Green Political Thought." 4th edition. London/New York: Routledge. DOI: 10.4324/9780203964620.

Dumaine, Carol; & Mintzer, Irving (2015). "Confronting Climate Change and Reframing Security." The SAIS Review of International Affairs. Vol. 35 (1). pp. 5–16. DOI: 10.1353/sais.2015.0014.

Dyer, Hugh (2017). "Green Theory." In: McGlinchey, Stephan; Walters, Rosie; & Scheinpflug, Christian (Eds.). International Relations Theory. E-International Relations Publishing, Bristol, Chapter 11. pp. 84–90.

Dzebo, Adis; Janetschek, Hannah; Brandi, Clara; & Iacobuta, Gabriela (2019). "Connections between the Paris Agreement and the 2030 Agenda: the case for policy coherence." SEI Working Paper. Stockholm Environment Institute, Stockholm. pp. 1–38.

Eckersley, Robyn (2012). "Moving Forward in the Climate Negotiations: Multilateralism or Minilateralism?" Global Environmental Politics. Vol. 12 (2), pp. 24–42. DOI:10.1162/GLEP_a_00107.

Eckersley, Robyn (2021). "Greening states and societies: from transitions to great transformations." Environmental Politics. Vol. 30 (1-2), pp. 245–265. DOI: 10.1080/09644016.2020.1810890.

Eckersley, Robyn (1993). "Free market environmentalism: Friend or foe?" Environmental Politics, Vol. 2 (1), pp. 1–19. DOI: 10.1080/0964 4019308414061.

Ecofys und Fraunhofer IWES (2017). "Smart-Market-Design in deutschen Verteilnetzen. Entwicklung und Bewertung von Smart Markets und Ableitung einer Regulatory Roadmap." Studie im Auftrag von Agora Energiewende. 110/02-S-2017/DE. 03/2017. Berlin. pp. 1–155.

Elkind, Jonathan (2010). "Energy Security: Call for a Broader Agenda." In: Pascual, Carlos; & Elkind, Jonathan (Editors). Energy Security, Economics, Politics, Strategies, and Implications, Washington: Brookings Institution Press, pp. 119–148.

Esakova, Nataliya (2012). "European Energy Security. Analyzing the EU-Russia Energy Security Regime in Terms of Interdependence Theory." Wiesbaden: Springer VS. DOI: 10.1007/978-3-531-19201-7.

European Commission (2023a). "REPowerEU: affordable, secure and sustainable energy for Europe." URL: https://commission.europa.eu/strategy-and-policy/priorities-2019-2024/european-green-deal/repowereu-affordable-secure-and-sustainable-energy-europe_en (accessed 10.05.2023),

European Commission (2023b). "A European Green Deal. Striving to be the first climate-neutral continent." URL: https://commission.europa.eu/strategy-and-policy/priorities-2019-2024/european-green-deal_en (accessed 10.05.2023).

European Commission (2023c). "Renewable energy targets." URL: https://energy.ec.europa.eu/topics/renewable-energy/renewable-energy-directive-targets-and-rules/renewable-energy-targets_en (accessed 10.05.2023).

European Commission (2023d). "Implementing and delegated acts - Taxonomy Regulation." URL: https://finance.ec.europa.eu/regulation-and-supervision/financial-services-legislation/implementing-and-delegated-acts/taxonomy-regulation_en (accessed 16.05.2023).

European Commission (2023e). "European Climate Law." URL: https://climate.ec.europa.eu/eu-action/european-green-deal/european-climate-law_en (accessed 4.06.2023).

European Commission (2022). "Communication from the Commission to the European Parliament, the Council, the European Economic and Social Committee and the Committee of the Regions. Digitalising the energy system – EU action plan." Strasbourg, 18.10.2022. URL: https://eur-lex.europa.eu/legal-content/EN/TXT/PDF/?uri=CELEX:52022DC0552&from=EN (accessed, 7.03.2023).

European Commission (2015). "Study on the effective integration of Distributed Energy Resources for providing flexibility to the electricity system." Final report to The European Commission 20. April 2015. URL: https://energy.ec.europa.eu/system/files/2015-06/5469759000%2520Effective%2520integration%2520of%2520DER%2520Final%2520ver%25202_6%2520April%25202015_0.pdf (reaccessed 30.05.2023).

EPRS (European Parliamentary Research Service) (2019). "Blockchain and the General Data Protection Regulation. Can distributed ledgers be squared with European data protection law?" Scientific Foresight Unit (STOA) PE 634.445 – July 2019. Brussels.

Eurostat (2022). "Nuclear Energy Statistics." URL: https://ec.europa.eu/eurostat/statistics-explained/index.php/Nuclear_energy_statistics (accessed 12.06.2020).

Federal Constitutional Court (2021). "Constitutional complaints against the Federal Climate Change Act partially successful." Press Release No. 31/2021 of 29 April. URL: https://www.bundesverfassungsgericht.de/SharedDocs/Pressemitteilungen/EN/2021/bvg21-031.html (accessed 7.03.2023).

Federal Energy Regulatory Commission (2007). "The Potential Benefits of Distributed Generation and Rate-Related Issues that May Impede its Expansion." A Study Pursuant to Section 1817 of the Energy Policy Act Of 2005. United States Department of Energy. URL: https://www.ferc.gov/sites/default/files/2020-04/1817_study_sep_07.pdf (accessed 4.07.2020).

Federal Parliament (Bundestag) (2007). „CO2-Bilanzen verschiedener Energieträger im Vergleich. Zur Klimafreundlichkeit von fossilen Energien, Kernenergie und erneuerbaren Energien." Wissenschaftlicher Dienste des Bundestages. pp. 1–8. URL: https://www.bundestag.de/resource/blob/406432/70f77c4c170d9048d88dcc3071b7721c/wd-8-056-07-pdf-data.pdf. (accessed 2.06.2020).

Fetting, Constanze (2020). "The European Green Deal." ESDN Report December 2020, ESDN Office, Vienna. pp. 1–22.

Financial Times (2020). "CO2 levels in atmosphere hit new highs despite coronavirus crisis." 4.06.2020. URL: https://www.ft.com/content/b3f309a3-ed87-474f-9ea0-d39bf4a5b246 (accessed 5.06.2020).

Fischer, Severin (2014). "Der neue EU-Rahmen für die Energie- und Klimapolitik bis 2030." Stiftung Wissenschaft und Politik. SWP-Aktuell, Vol. 73, pp. 1–8.

Flaute, Markus; Großmann, Anett; Lutz, Christian; & Nieters, Anne (2017). "Macroeconomic Effects of Prosumer Households in Germany." International Journal of Energy Economics and Policy. 2017, 7(1), pp. 146–155.

Florides, Georgios A.; & Christodoulides, Paul (2008). "Global warming and carbon dioxide through sciences." Environmental International. Vol. 35 (2). pp. 390–401. DOI: 10.1016/j.envint.2008.07.007

Fraunhofer FOKUS (no date). "WindNODE." URL: https://www.fokus.fraunhofer.de/en/sqc/projects/windnode (accessed 13.09.2020).

Freier, Alexander (2022a). "Blockchain in the energy sector. An analysis of the Brooklyn case." SSRN: https://papers.ssrn.com/sol3/papers.cfm?abstract_id=3998651 (posted on 5.01.2022).

Freier, Alexander (2022b). "Digitalization and flexibility trading in the energy sector. Lessons learned from Northeastern Germany." SSRN. https://papers.ssrn.com/sol3/papers.cfm?abstract_id=3998659 (posted 5.01.2022).

Frizzo-Barker, Julie; Chow-White, Peter A.; Adams, Phillipa R.; Mentanko, Jennifer; Ha, Dung; & Green, Sandy (2020). "Blockchain as a disruptive technology for business: A systematic review." International Journal of Information Management, pp. 1–14. DOI: 10.1016/j.ijinfomgt.2019.10.014.

Frøystad, Peter; & Holm, Jarle (2015). "Blockchain. Powering the Internet of Value." Whitepaper EVRY Financial Services. pp. 150. URL: https://blockchainlab.com/pdf/bank-2020---blockchain-powering-the-internet-of-value---whitepaper.pdf. (accessed 15.05.2020).

Gährs, Swantje; Aretz, Astrid; Flaute, Markus; Oberst, Christian A; Großmann, Anett; Lutz, Christian; Bargende, Daniel; Hirschl, Brend; & Madlener, Reinhard (2016). "Prosumer-Haushalte: Handlungsempfehlungen für eine sozial-ökologische und systemdienliche Förderpolitik," Aachen URL: <https://www.prosumer-haushalte.de/data/prohaus/user_upload/Dateien/Prosumer-Haushalte__Handlungsempfehlungen.pdf> (accessed 13.01.2020).

Gandenberger, Carsten (2015). "Theoretical Perspectives on the International Transfer and Diffusion of Climate Technologies." Working Paper Sustainability and Innovation. No. S 12/2015. Fraunhofer Institute. pp. 1–38.

Gassmann, Oliver; Frankenberger, Karolin; & Csik, Michaela (2014). "Revolutionizing the Business Model." In: Gassmann, Oliver & Schweitzer, Fiona (Editors). Management of the Fuzzy Front End of Innovation. Cham/Heidelberg: Springer, pp. 89–97. DOI: 10.1007/978-3-319-01056-4_7.

Gerlagh, Reyer; Heijmans, Roweno J. R. K.; & Rosendahl, Knut Einer (2022). "Shifting concerns for the EU ETS: are carbon prices becoming too high?" Environmental Research Letters. Vol. 17 (054018). pp. 1–6. DOI: 10.1088/1748-9326/ac63d6.

German Federal Environmental Agency (2023). "Erneuerbare Energien in Deutschland. Daten zur Entwicklung im Jahr 2022." Hintergrund. February 2023. pp. 1–28.

German Federal Government (2021). "Climate Change Act 2021. Intergenerational contract for the climate." URL: https://www.bundesregierung.de/breg-de/themen/klimaschutz/climate-change-act-2021-1913970 (accessed 7.03.2023).

Ghiani, Emilio; Pilo, Fabrizio; & Celli, Gianni (2018). "Definition of Smart Distribution Networks." In: Zare, Kazem; & Nojavan, Sayyad (Editors). Operation of Distributed Energy Resources in Smart Distribution. Elsevier: London., pp. 1–24. DOI10.1016/B978-0-12-814891-4.00001-1.

GIZ (2019a). "Blockchain meets Energy. Digital Solutions for a Decentralized and Decarbonized Sector." German-Mexican Energy Partnership (EP) and Florence School of Regulation (FSR). p. 1–41.

GIZ (2019b). "Blockchain. A World without Middlemen. Promise and Practise of Distributed Governance." Deutsche Gesellschaft für Internationale Zusammenarbeit (GIZ) GmbH Blockchain Lab. Berlin/Eschborn. pp. 1–92.

Gómes, Oscar A.; & Gasper, Des (2013). „Human Security. A Thematic Guidance Note for Regional and National Human Development Report Teams." United Nations Development Programme. Human Development Report Office. pp. 1–16.

Gough, Matthew; Castro, Rui; Santos, Sergio F.; Shafie-khah, Miadreza; & Catalão, João P.S. (2020). "A panorama of applications of blockchain technology to energy." In: Shafie-khah, Miadreza (Editor). Blockchain-based Smart Grids. London: Elsevier/Academic Press. pp. 5–41. DOI: https://doi.org/10.1016/B978-0-12-817862-1.00002-6.

Grin, John; & Loeber, Anne (2007). "Theories of policy learning: Agency, structure and change." In: Fischer, Frank; Miller, Gerald J; Sidney, Mara S. (Editors). Handbook of Public Policy Analysis. Theory, Politics, and Methods. London/New York: Taylor & Francis. pp. 201–219.

Grüner, Andreas; Mühle, Alexander; Gayvoronskaya, Tatiana; & Christoph, Meinel (2018). "A Quantifiable Trust Model for Blockchain-based Identity Management." 2018 IEEE International Conference on Internet of Things (iThings) and IEEE Green Computing and Communications (GreenCom) and IEEE Cyber, Physical and Social Computing (CPSCom) and IEEE Smart Data (SmartData). pp. 1475–1482. DOI: 10.1109/Cybermatics_2018.2018.00250.

Guo, Yuanxiong; Fang, Yuguang; & Khargonekar, Pramod P. (2017). "Stochastic Optimization for Distributed Energy Resources in Smart Grids." Cham: Springer. DOI: 10.1007/978-3-319-59529-0.

Hafid, Abdelatif; Hafid, Abdelhakim Senhaji; Samih; & Samih, Mustapha (2020). "Scaling Blockchains: A Comprehensive Survey." IEEE Access, pp. 125244–125262. DOI: 10.1109/ACCESS.2020.3007251.

Hall, Stephan; Brown, Donal; Davis, Mark; Ehrtmann, Moritz; & Holstenkamp, Lars (2020). "Business Models for Prosumers in Europe." PROSEU - Prosumers for the Energy Union: Mainstreaming active participation of citizens in the energy transition (Deliverable N°D4.1). pp. 1–92. URL: https://proseu.eu/sites/default/files/Resources/PROSEU_D4.1_Business%20models%20for%20collective%20prosumers.pdf (accessed 17.05.2020).

Hall, Rodney Bruce; & Biersteker, Thomas J. (2004). "The emergence of private authority in the international system." In: Hall, Rodney Bruce & Biersteker, Thomas J. (editors). The Emergence of Private Authority in Global Governance. Cambridge: Cambridge University Press. pp. 3–22. DOI: 10.1017/CBO9780511491238.002.

Hassan, Samer; & De Filippi, Primavera (2021). "Decentralized Autonomous Organization." Internet Policy Review, Vol. 10(2). pp. 1–10. DOI: 10.14763/2021.2.1556.

Hatziargyriou, Nikos D.; Jenkins, Nick; Strbac, G.; Peças Lopes, João Abel; Ruela, José; Engler, Alfred; Kariniotakis, George; Oyarzabal, José; & Amorim, António (2006). "Microgrids-Large Scale Integration of Microgeneration to Low Voltage Grids." Proceedings of CIGRE 2006, 41st annual session conference. pp. 1–12.

Hawlitschek, Florian; Notheisen, Benedikt; & Teubner, Timm (2018). "The limits of trust-free systems: A literature review on blockchain technology and trust in the sharing economy". Electronic Commerce Research and Applications, Vol. 29, pp. 50–63. DOI: 10.1016/j.elerap.2018.03.005.

Herzog, Maximilian (2022). "The future of coal in Europe: is this the exit from the exit?" Friedrich Ebert Stiftung. 12.10.2022. URL: https://justclimate.fes.de/e/the-future-of-coal-in-europe (accessed. 15.04.2023).

Hilal, Ala' Abu; Badra, Mohamad; & Tubaishat, Abdallah (2022). "Building Smart Contracts for COVID-19 Pandemic over the Blockchain Emerging Technologies." Procedia Computer Science 198 (2022). pp. 323–328. DOI: 10.1016/j.procs.2021.12.248.

Hira, Anil; & Cohn, Theodore H. (2003). "Toward a Theory of Global Regime Governance." International Journal of Political Economy, vol. 33, no. 4, Winter 2003–4, pp. 4–27. DOI: 10.1080/08911916.2003.11042909.

Hitschler, Werner & Kellermann, Dieter (2020). "DT:HUB – Etablierte und Start-ups machen Zukunft." In: Doleski, Oliver D. (editor). Realisierung Utility 4.0, Vol. 1. Praxis der digitalen Energiewirtschaft von den Grundlagen bis zur Verteilung im Smart Grid. Wiesbaden: Springer Vieweg, pp. 131–140. DOI: 10.1007/978-3-658-25332-5_7.

Hoffman, Andrew J.; & Jennings, P. Devereaux (2015). "Institutional Theory and the Natural Environment." Organization & Environment. Vol. 28 (1). Special Issue: Review of the Literature on Organizations and Natural Environment: From the Past to the Future. pp. 8–31. DOI: 10.1177/1086026615575331.

Horn, Miriam; & Mirzatuny, Marita (2013). "Mining Big Data to Transform Electricity." In: Noam, Eli M.; Pupillo, Lorenzo Maria; Kranz, Johann J. (Editors). Broadband Networks, Smart Grids and Climate Change. New York: Springer. pp. 47–58. DOI: 10.1007/978-1-4614-5266-9_6.

IAEA (2023). "Power Reactor Information System – Germany." URL: https://pris.iaea.org/PRIS/CountryStatistics/CountryDetails.aspx?current=DE (accessed 6.04.2023).

Ibañez, Juan Ignacio; & Freier, Alexander (2023). "Bitcoin's Carbon Footprint Revisited: Proof of Work Mining for Renewable Energy Expansion." Challenges. Vol.14(3), 35. pp. 1 – 21. DOI: 10.3390/challe14030035.

IEA (2019). "Global Energy & CO2 Status Report 2019." IEA, Paris. URL: https://www.iea.org/reports/global-energy-co2-status-report-2019.

IPCC (2023). "Synthesis report of the IPCC Sixth Assessment Report (AR6)." Summary for Policymakers. URL: https://report.ipcc.ch/ar6syr/pdf/IPCC_AR6_SYR_SPM.pdf (accessed 06.04.2023).

IPCC (2018). "Global warming of 1.5°C. An IPCC Special Report on the impacts of global warming of 1.5°C above pre-industrial levels and related global greenhouse gas emission pathways, in the context of strengthening the global response to the threat of climate change, sustainable development, and efforts to eradicate poverty." (Masson-Delmotte, Valérie; Zhai, Panmao; Pörtner, Hans-Otto; Roberts, Debra; Skea, Jim; & Shukla, Priyadarshi R.; Pirani, Anna; Moufouma-Okia, Wilfran; Péan, Clotilde; Pidcock, Roz; Connors, Sarah; Matthews, J. B. Robin; Chen, Yang; Zhou, Xiao; Gomis, Melissa I.; Lonnoy, Elisabeth; Maycock, Tom; Tignor; Melinda; & Waterfield, Tim). URL: https://www.ipcc.ch/site/assets/uploads/sites/2/2019/06/SR15_Full_Report_High_Res.pdf (accessed 5.05.2020).

IPCC (2014). "Climate Change 2014 Synthesis Report." Synthesis Report Summary for Policymakers. URL: https://www.ipcc.ch/site/assets/uploads/2018/02/AR5_SYR_FINAL_SPM.pdf (accessed 5.05.2020).

IPCC (2012). "Renewable Energy Sources and Climate Change Mitigation Summary for Policymakers and Technical Summary." URL: https://archive.ipcc.ch/pdf/special-reports/srren/SRREN_FD_SPM_final.pdf (accessed 8.06.2020).

IRENA (2023). "Renewable capacity statistics 2023. International Renewable Energy Agency." Abu Dhabi. URL: https://mc-cd8320d4-36a1-40ac-83cc-3389-cdn-endpoint.azureedge.net/-/media/Files/IRENA/Agency/Publication/2023/Mar/IRENA_RE_Capacity_Statistics_2023.pdf?rev=d2949151ee6a4625b65c82881403c2a7 (accessed 13.04.2023).

IRENA (2020). "Innovation landscape brief: Co-operation between transmission and distribution system operators." International Renewable Energy Agency. Abu Dhabi.

IRENA (2019). "Climate Change and Renewable Energy. National Policies and the Role of Communities, Cities, and Regions." A report from the International Renewable Energy Agency (IRENA) to the G20 Climate Sustainability Working Group (CSWG), pp. 1–60.

IRENA (2017). "Renewable Energy. A Key Climate Solution." URL: https://www.irena.org/-/media/Files/IRENA/Agency/Publication/2017/Nov/IRENA_A_key_climate_solution_2017.pdf?la=en&hash=A9561C1518629886361D12EFA11A051E004C5C98 (accessed 8.7.2020).

IRENA (2015). "Smart Grids and Renewables. A Cost-Benefit Analysis Guide for Developing Countries." URL: https://www.irena.org/-/media/Files/IRENA/Agency/Publication/2015/IRENA_PST_Smart_Grids_CBA_Guide_2015.pdf (accessed 20.05.2020)

IRENA (2013). "Smart Grids and Renewables. A Guide for Effective Deployment." Working Paper, pp. 1–44.

Islam, S. Nazrul; & Winkel, John (2017). "Climate Change and Social Inequality." DESA Working Paper. No. 152, pp. 1–30.

Jäger-Waldau, Arnulf (2019). "PV Status Report 2019." JRC Science for Policy Report. Joint Research Centre. Luxemburg: Publications Office of the European Union. DOI: 10.2760/326629.

Jarvis, Stephen; Deschenes, Olivier; & Jha, Akshaya (2022). "The Private and External Costs of Germany's Nuclear Phase-Out." Journal of European Economic Association. Vol. 20(3), pp. 1311–1346. DOI: 10.1093/jeea/jvac007.

Kalla, Anshuman; Hewa, Tharaka; Mishra, Raaj Anand; Ylianttila, Mika, & Liyanage, Madhusanka (2020). "The Role of Blockchain to Fight Against COVID-19." IEEE Engineering Management Review. pp. 1–10. DOI: 10.1109/EMR.2020.3014052.

Kirli, Desen; Couraud, Benoit; Robu, Valentin; Salgado-Bravo, Marcelo; Norbu, Sonam; Andoni, Merlinda; Antonopoulos, Ioannis; Negrete-Pincetic, Matias; Flynn, David; & Kiprakis, Aristidis (2022). "Smart contracts in energy systems: A systematic review of fundamental approaches and implementations." Renewable and Sustainable Energy Reviews. Vol. 158 (112013). pp. 1–28. DOI: 10.1016/j.rser.2021.112013.

Kirstein, Fabian; Lämmel, Philipp; & Altenbernd, Anton (2021). "Mythos Blockchain. Zwischen Hoffnung und Realität. Kompetenzzentrum Öffentliche IT." Fraunhofer-Institut für Offene Kommunikationssysteme (FOKUS) & Weizenbaum-Institut. Berlin.

Krasner, Stephen D. (1983). "Structural causes and regime consequences: regimes as intervening variables." In: (ibid.) International Regimes. Ithaca/London: Cornell University Press. pp. 1–21.

Keohane, Robert O.; & Oppenheimer, Michael (2016). "Paris: Beyond the Climate Dead End through Pledge and Review?" Politics and Governance, Vol.4 (3), pp. 142–151. DOI: https://doi.org/10.17645/pag.v4i3.634.

Keohane, Robert O.; & Victor, David G. (2015). "After the failure of the top-down mandates. The role of experimental governance in climate change policy." In: Barrett, Scott; Carraro, Carlo; & De Melo, Jamie (editor). Towards a Workable and Effective Climate Change Regime. London: CEPR Press, pp. 201–212.

Keohane, Robert O.; & Victor, David G. (2011). "Regime Complex for Climate Change." Perspectives on Politics. Vol. 9(1). pp. 7–23. DOI: 10.1017/S1537592710004068.

Keohane, Robert O. (1984). "After Hegemony. Cooperation and Discord in the World Political Economy." Princeton: Princeton University Press.

Keohane, Robert O. (1982). „The Demand for International Regimes." International Organization. Vol. 36(2). pp. 325–355. DOI: 10.1017/S00 2081830001897X.

Kirsten, Selder (2014). „Renewable Energy Sources Act and Trading of Emission Certificates: A national and a supranational tool direct energy turnover to renewable electricity-supply in Germany." Energy Policy. Vol. 64, pp. 302–312. DOI: 10.1016/j.enpol.2013.08.030.

Knorr, Kaspar; Schütt, Jonathan; Strahlhoff, Julia; Kroschewski, Theresa; Siegl, Stefan; Werner, Uta; Willner, Alexander; Eckert, Klaus-Peter; Wolf, Armin; Lämmel, Philipp; & Haase, Peter (2019). "Weiße Flecken in der digitalen Vernetzung der Energiewirtschaft. IUK-Ansätze und Lösungen." Fraunhofer Institut. Fraunhofer-Institut für Energiewirtschaft und Energiesystemtechnik (IEE) und Fraunhofer-Institut für offene Kommunikationssysteme (FOKUS). Berlin. pp. 1–74. DOI: 10.24406/publica-fhg-299909.

Koirala, Binod Prasad; Hakvoort, Rudi A.; Van Oost, Ellen C.; & van der Windt, Henny J. (2019). „Community Energy Storage: Governance and Business Models." In: Sioshansi, Fereidoon (editor). Consumer, Prosumer, Prosumager. How Service Innovations Will Disrupt the Utility Business Model. Academic Press: London. pp. 209–234. DOI: 10.1016/B978-0-12-816835-6.00010.

Kung, Chih-Chun; & McCarl, Bruce A. (2018). "Sustainable Energy Development under Climate Change." Sustainability 2018, Vol. 10(9), pp. 1–4. DOI: https://doi.org/10.3390/su10093269.

Kurth, Matthias (2013). "Smart Metering, Smart Grids, Smart Market Design." In: Noam, Eli M.; Pupillo, Lorenzo Maria; & Kranz, Johann J. (Editors). Broadband Networks, Smart Grids, and Climate Change. New York: Springer, pp. 11–15. DOI: 10.1007/978-1-4614-5266-9_2.

Laing, Tim; Sato, Misato; Grubb, Michael; & Comberti, Claudia (2013). "Assessing the effectiveness of the EU emissions trading system." Centre for Climate Change Economics and Policy. Working Paper No.126, pp. 1–35. https://www.lse.ac.uk/granthaminstitute/wp-content/uploads/2014/02/WP106-effectiveness-eu-emissions-trading-system.pdf.

Le Fevre, Chris (2017). "Methane Emissions. From blond spot to spotlight." The Oxford Institute for Energy Studies. OIES Paper: NG 122, pp. 1–39.

Lejano, Raul P.; & Stokols, Daniel (2013). "Social ecology, sustainability, and economics." Ecological Economics. Vol. 89. pp. 1–6. DOI: 10.1016/j.ecolecon.2013.01.011.

Le Quéré, Corinne; Jackson, Robert B., Jones, Matthew W.; Smith, Adam J.P.; Abernethy, Sam; Andrew, Robbiwe M. De-Gol, Anthony J.; Willis, David R.; Shan, Yuli; Canadell, Joseph G; Friedlingstein, Pierre; Creutzig, Felix; & Peters, Glen P. (2020). "Temporary reduction in daily global CO2 emissions during the COVID-19 forced confinement." Nature Climate Change. Vol. 10. pp. 647–653. DOI: 10.1038/s41558-020-0797-x.

Li, Rita Yi Man (2018). "An Economic Analysis on Automated Construction Safety. Internet of Things, Artificial Intelligence and 3D Printing." Singapore: Springer. DOI: https://doi.org/10.1007/978-981-10-5771-7.

Liu, Zhu; Ciais, Philippe; Deng, Zhu; Lei, Ruixue; Davis, Steven J.; Feng, Sha; Zheng, Bo; Cui, Duo; Dou, Xinyu; Zhu, Biqing; Guo, Rui; Ke, Piyu; Sun, Taochun; Lu, Chenxi; He, Pan; Wang, Yuan; Yue, Xu; Wang, Yilong; Lei, Yadong; Zhou, Hao; Cai, Zhaonan; Wu, Yuhui; Guo, Runtao; Han, Tingxuan; Xue, Jinjun; Boucher, Olivier; Boucher, Eulalie; Chevallier, Frédéric; Tanaka, Katsumasa; Wei, Yiming; Zhong, Haiwang; Kang, Chongqing; Zhang, Ning; Chen, Bin; Xi, Fengming; Liu, Miaomiao; Bréon, François-Marie; Lu, Yonglong Lu; Zhang, Qiang; Guan, Dabo; Gong, Peng; Kammen, Daniel M.; He, Kebin; & Schellnhuber, Hans Joachim (2020). "Near-real-time monitoring of global CO2 emissions reveals the effects of the COVID-19 pandemic." Nature Communications. Vol. 11 (5172). pp. 1–12. DOI: 10.1038/s41467-020-18922-7.

Lopes, João Abel Peças; Madureira, André; Gil, Nuno; & Resende, Fernanda (2014). "Operation of Multi-Microgrids." In: Hatziargyriou. Nikos (Editor). Microgrids: Architectures and Control, West Sussex: John Wiley & Sons Publishing, pp. 165–205. DOI: 10.1002/9781118720677.ch05.

Lowitzsch, Jens; Hoicka, Christina E.; & van Tulder, Felicia (2020). "Renewable energy communities under the 2019 European Clean Energy Package—Governance model for the energy clusters of the future?" Renewable and Sustainable Energy Reviews. Vol. 122, pp. 1–13. DOI: 10.1016/j.rser.2019.109489.

Lowitzsch, Jens (2019). "Investing in a Renewable Future—Renewable Energy Communities, Consumer (Co-) Ownership and Energy Sharing in the Clean Energy Package." Renewable Energy Law and Policy. Vol. 9(2), pp. 14–36. DOI: 10.4337/relp.2019.02.02.

Lowitzsch, Jens & Hanke, Florian (2019). "Consumer (Co-)ownership in Renewables, Energy Efficiency and the Fight Against Energy Poverty—a Dilemma of Energy Transition." Renewable Energy Law and Policy Review. Vol. 9(3). pp.5–21. DOI: 10.4337/relp.2019.03.01.

Lv, Tianguang; & Ai, Qian (2016). "Interactive energy management of networked microgrids-based active distribution system considering large-scale integration of renewable energy resources." Applied Energy. Vol. 163, 408–422. DOI: 10.1016/j.apenergy.2015.10.179.

Madhumathi, T. K.; Nagadeepa, C.; Snehavalli, R. (2021). "Change Of Work Culture During Covid-19." International Journal of Aquatic Biology. Vol. 12(02), pp. 308–314.

Marczinkowski, Hannah Mareike; Alberg Østergaard, Poul; & Roth Djørup, Søren (2019). "Transitioning Island Energy Systems–Local Conditions, Development Phases, and Renewable Energy Integration." Energies. Vol. 12(3484). pp. 3–20. DOI: 10.3390/en12183484.

Mason, Michael (2013). "Climate change and human security. The international governance architectures, policies, and instruments." In: Redclift, Michael R. & Grasso, Marco (2013). Handbook on Climate Change and Human Security. Cheltenham/Northampton: Edward Elgar, pp. 382–401.

Mattila, Juri (2016). "The Blockchain Phenomenon. The Disruptive Potential of Distributed Consensus Architectures." Berkeley Roundtable of the International Economy (BRIE) Working Paper 2016-1. University of California, Berkeley. pp.1–25. https://www.researchgate.net/publication/313477689_The_Blockchain_Phenomenon_-The_Disruptive_Potential_of_Distributed_Consensus_Architectures. (accessed 3.05.2020).

Mayntz, Renate (2002). "National States and Global Governance." VII Inter-American Congress of CLAD on State and Public Administration Reform Lisbon, Portugal, October 8-11, 2002. pp. 1–8.

Meinel, Christoph; Gayvoronskaya, Tatiana; & Schnjakin, Maxim (2018). "Blockchain. Hype oder Innovation." Technische Berichte Nr. 113 des Hasso-Plattner-Instituts für Digital Engineering an der Universität Potsdam, Potsdam. pp.1–116.

Meng, T., Y. Zhao, K. Wolter, and C-Z. Xu (2021). "On Consortium Blockchain Consistency: A Queueing Network Model Approach." IEEE Transactions on Parallel and Distributed Systems, Vol. 32(6). pp. 1369–1382. DOI: 10.1109/TPDS.2021.3049915.

Mengelkamp, Esther; Gärttner, Johannes; Rock, Kerstin; Kessler, Scott; Orsini, Laurence; & Weinhardt, Christof (2018). "Designing microgrid energy markets. A case study: The Brooklyn Microgrid." Applied Energy. Vol. 210 (2018). pp. 870–880. DOI: 10.1016/j.apenergy.2017.06.054.

Menner, Martin; & Götz, Reichert (2020). "EU Hydrogen Strategy." In: Centrum für Europäische Politik. cepPolicyBrief No. 2020-14. pp. 1–4.

Mercure, J.-F.; Pollitt, H; Chewpreecha, U.; Salas, P.; Foley, A.M.; Holden, P.M.; & Edwards, M.R. (2014). "The dynamics of technology diffusion and the impacts of climate policy instruments in the decarbonisation of the global electricity sector." Energy Policy. Vol. 73, pp. 686–700. DOI: 10.1016/j.enpol.2014.06.029.

Mika, Bartek; & Goudz, Alexander (2020). "Blockchain-Technologie in der Energiewirtschaft. Blockchain als Treiber der Energiewende." Springer Vieweg: Bottrop. DOI: 10.1007/978-3-662-60568-4.

Minniti, Simone; Haque, Niyam; Nguyen, Phuong; & Pemen, Guus (2018). "Local Markets for Flexibility Trading: Key Stages and Enablers." Energies. Vol. 11(11), 3074. pp. 1–21. DOI: 10.3390/en11113074.

Moomaw, William; Maurice, Lourdes; Yamba, Francis Davison; & Jäger-Waldau (2012). "Renewable Energy and Climate Change." In: Edenhofer, O.; Pichs-Madruga, R.; Sokona, Y; Seyboth, K.; Matschoss, P.; Kadner, S.; Zwickel, T.; Eickemeier, P.; Hansen, G; Schlömer, S.; von Stechow, C. (Editors). IPCC Special Report on Renewable Energy Sources and Climate Change Mitigation. Cambridge: Cambridge University Press. pp. 161–208. DOI: 10.1017/CBO9781139151153.005.

Munzel, Benjamin; Reiser, Marco; & Steinbacher, Karoline (2022). "Flexibilitätspotenziale und Sektorkopplung." Synthesebericht 1 des SINTEG Förderprogramms. Studie im Auftrag des BMWK. Berlin.

Nanda, Nitya; & Srivastava, Nidhi (2009). "Clean Technology Transfer and Intellectual Property Rights." Sustainable Development Law & Policy. Vol. 9(3). Spring 2009: Clean Technology and International Trade. pp. 42–46, 68–69.

Nakamoto, Satoshi (2009). "Bitcoin: A Peer-to-Peer Electronic Cash System." White Paper. 1–9. URL: https://bitcoin.org/bitcoin.pdf (accessed 25.09.2020).

Neubauer, Maik (2020). "Das Europaïsche Hochspannungsnetz—Die Zukunft von Big Data und künstlicher Intelligenz in kritischen Infrastrukturen." In: Doleski, Oliver D. (Editor). Realisierung Utility 4.0. Vol. 1. Praxis der digitalen Energiewirtschaft von den Grundlagen bis zur Verteilung im Smart Grid. Wiesbaden: Springer Vieweg. pp. 723–737. DOI: 10.1007/978-3-658-25332-5_44.

New York City Neighborhood Tabulation Areas (2010). "Table PL-P5 NTA: Total Population and Persons Per Acre." URL: https://www1.nyc.gov/assets/planning/download/pdf/data-maps/nyc-population/census2010/t_pl_p5_nta.pdf (accessed 30.08.2020).

Nieße, Astrid; Lehnhoff, Sebastian; Troschel, Martin; Uslar, Matthias; Wissing, Carsten; Appelrath, H.-Jürgen; & Sonnenschein, Michael. (2012). "Market-based self-organized provision of active power and ancillary services: An agent-based approach for Smart Distribution Grids." 2012 Complexity in Engineering (COMPENG). Proceedings. pp. 1–5. DOI: 10.1109/CompEng.2012.6242953.

Nocentini, Stefano; Gavazzi, Roberto; & Pupillo, Lorenzo Maria (2013). "Broadband ICT and Smart Grids: A Win-Win Approach." In: Noam, Eli M.; Pupillo, Lorenzo Maria; Kranz, Johann J. (Editors). Broadband Networks, Smart Grids and Climate Change. New York: Springer, pp. 17–31. DOI: 10.1007/978-1-4614-5266-9_3.

Nojavan, Sayyad; & Zare, Kazem (2020). "Preface." In: Nojavan, Sayyad; & Zare, Kazem (Editors). Demand Response Application in Smart Grids. Operation Issues. Vol. 2, Cham: Springer. pp. v–vi. DOI: 10.1007/978-3-030-32104-8.

OECD (2019). "Blockchain technologies as a digital enabler for sustainable infrastructure." OECD Environment Policy Papers, No. 16, OECD Publishing, Paris.

Opp, Karl-Dieter (2002). "Methodologie der Sozialwissenschaften. Einführung in Probleme ihrer Theoriebildung und praktischen Anwendung." Wiesbaden: Westdeutscher Verlag. DOI: 10.1007/978-3-322-95673-6.

Orsini, Lawrence; Kessler, Scott; Wei, Julianna; & Field, Heather (2019). "How the Brooklyn Microgrid and TransActive Grid are paving the way to next-gen energy markets." In: Su, Wencong; & Huang, Alex Q. (Editors) The Energy Internet. An Open Energy Platform to Transform Legacy Power Systems into Open Innovation and Global Economic Engines. Duxford/Cambridge: Woodhead Publishing, pp. 223–239. DOI: 10.1016/B978-0-08-102207-8.00010-2.

Panahi, Farzad H.; & Panahi, Fereidoun H. (2020). "Smart Grids and Green Wireless Communications" (Chapter 1) In: Nojavan, Sayyad; & Zare, Kazem (Editors). Demand Response Application in Smart Grids. Operation Issues. Vol. 2, Cham: Springer. pp. 1–35. DOI: 10.1007/978-3-030-32104-8_1.

Parag, Yael; & Sovacool, Benjamin K. (2016). "Electricity market design for the prosumer era." Nature Energy, Vol. 1, pp. 1–6. DOI: 10.1038/nenergy.2016.32.

Parizi, Reza M.; Amritraj; & Dehghantanha, Ali (2018). "Smart Contract Programming Languages on Blockchains: An Empirical Evaluation of Usability and Security." Blockchain – ICBC 2018., Proceedings. pp. 75–91. DOI: 10.1007/978-3-319-94478-4_6.

Paterson, Matthew (2022). "Green Theory." In: Devetak, Richard; & True, Jacqui (Editors). Theories of International Relations. 6th Edition, London/New York: Bloomsbury Academic., pp. 462–503.

Paterson, Matthew (2005). "Green Politics." In: Burchill, Scott; Linklater, Andrew; Devetak, Richard; Donnelly, Jack; Paterson, Matthew; Reus-Smit, Christian & True, Jacqui (editors). Theories of International Relations. 3rd edition. Houndmills: Palgrave Macmillan, pp. 235–257.

Powering Past Coal Alliance (2019). "Germany." URL: https://poweringpastcoal.org/members/germany/ (accessed 1.06.2023).

Qi, Tianyu; Zhang, Xiliang; & Karplus, Valerie J. (2014). "The energy and CO2 emissions impact of renewable energy development in China." Energy Policy. Vol. 68 (2014). pp. 60–69. DOI: 10.1016/j.enpol.2013.12.035.

Radtke, Jörg; Canzler, Weert; Schreurs, Miranda; & Wurster, Stefan (2018). "Die Energiewende in Deutschland—zwischen Partizipationschancen und Verflechtungsfalle." In: Radtke, Jörg; Kersting, Norbert (Editors). Energiewende. Politikwissenschaftliche Perspektiven. Wiesbaden: Springer VS., pp. 17–44. DOI: 10.1007/978-3-658-21561-3_2.

Rai, Neha; & Nash, Erin J. (2017). "Using political economy analysis as a tool in national planning." In: Rai, Neha; Fisher, Susannah (Editors). The Political Economy of Low Carbon Resilient Development. Routledge: London/New York., pp. 131–151.

Rising, James; Tedesco, Marco; Piontek, Franziska; & Stainforth, Davis A. (2022). "The missing risks of climate change." Nature. Vol. 610, pp. 643–651. DOI: 10.1038/s41586-022-05243-6.

Rogers, Everett M.; Singhal, Arvind; & Quinlan, Margaret (2009). "Diffusion of Innovations." In: Stacks, Don W.; Salwen, Michael B. (Editors). An Integrated Approach to Communication Theory and Research, Mahway, MJ: Lawrence Erlbaum Associates, pp. 418–434.

Rogers, Everett M. (2003). "Diffusion of Innovations." 5th edition. New York/London: Free Press.

Rosenau, James N.; & Czempiel, Ernst-Otto (1992). "Governance Without Government: Order and Change in World Politics." Cambridge: Cambridge University Press. DOI: 10.1017/CBO9780511521775.

Savić, Andrijana; & Dobrijević, Gordana (2022). "The impact of the Covid-19 pandemic on work organization." European Journal of Applied Economics. EJAE 2022, Vol. 19(1). pp. 1–15. DOI: 10.5937/EJAE19-35904.

Schill, Wolf-Peter; Zerrahn, Alexander; & Kunz, Friedrich (2017). "Prosumage of Solar Electricity: Pros, Cons, and the System Perspective." Economics of Energy & Environmental Policy. Vol. 6 (1). pp. 7–31. DOI: 10.5547/2160-5890.6.1.wsch.

Schnell, Rainer; Hill, Paul B.; & Esser, Elke (1999). "Methoden der empirischen Sozialforschung." 6. Auflage. München: Oldenbourg.

Schwaegerl, Christine; & Tao, Liang (2014). "The Microgrids Concept." In: Hatziargyriou, Nikos (editor). Microgrids Architectures and Control. West Sussex: Wiley. p. 1–24.

Shafie-khah, Miadreza (2020). "Blockchain-based Smart Grids." London: Elsevier/Academic Press.

Shah, Het; Shah, Manasi; Tanwar, Sudeep; & Kumar, Neeraj (2021). "Blockchain for COVID-19: a comprehensive review." Personal and Ubiquitous Computing. pp. 1–28. DOI: 10.1007/s00779-021-01610-8.

Sharma, Rakesh (2019). "Brooklyn Microgrid Gets Approval for Blockchain-based Energy Trading." EnergyCentral online. 26.12.2019. URL: https://energycentral.com/c/iu/brooklyn-microgrid-gets-approval-blockchain-based-energy-trading?utm_medium=PANTHEON_STRIPPED#ece-comments (accessed 8.05.2023).

Sinn, Hans-Werner (2017). "Buffering volatility: A study on the limits of Germany's energy revolution." European Economic Review. Vol. 99 (2017). pp. 130–150. DOI: 10.1016/j.euroecorev.2017.05.007.

Sinn, Hans-Werner (2012). "The Green Paradox. A Supply-Side Approach to Global Warming." Cambridge: The MIT PRESS.

Soeder, Daniel J. (2021). "Fossil Fuels and Climate Change." In: (Idem). Fracking and the environment. A scientific assessment of the environmental risks from hydraulic fracturing and fossil fuels. Cham: Springer International Publishing. pp. 155–186. DOI: 10.1007/978-3-030-59121-2_9.

Strange, Susan (1996). "The retreat of the state. The diffusion of power in the world economy." Cambridge: Cambridge University Press.

Sunyaev, Ali (2020). "Blockchain—'Like a Locked Train.'" In: Kühne, Christian (Editor). Blockchain Technology for Industrial Production and the Digital Circular Economy. Think Tank for Industrial Resource Strategies. pp. 16–27. URL: https://www.thinktank-irs.de/wp-content/uploads/2021/02/RZ_THINKTANK_Brochure_Blockchain_Druck_ENG_WEB.pdf. (accessed 9.12.2021)

Solomon, Susan; Plattner, Gian-Kasper; Knutti, Reto; & Friedlingstein, Pierre (2009). "Irreversible climate change due to carbon dioxide emissions." PNAS. Vol. 106 (6). pp. 1704–1709. DOI: 10.1073/pnas.0812721106.

Sustainable Energy for All (2014). "Sustainable Energy for All 2013-2014: Global Tracking Framework." International Bank for Reconstruction and Development and The World Bank 2014. Washington.

Steen, Marc Pierr E. Gregoire (2001). "Greenhouse gas emissions from fossil fuel fired power generation systems." European Commission. Joint Research Centre (Dg Jrc) Institute for Advanced Materials. pp. 1-61.

Steinbacher, Karoline (2019). "Exporting the Energiewende. German Renewable Energy Leadership and Policy Transfer." Wiesbaden: Springer VS. DOI: https://doi.org/10.1007/978-3-658-22496-7.

Stehr, Nico; & von Storch, Hans (1995). "The social construct of climate and climate change." Climate Research. Vol. 5(2), pp. 99–105. DOI: 10.3354/cr005099.

Stone, Diane (2012). "Transfer and translation of policy." Policy Studies, Vol. 33(6), pp. 483–499. DOI: https://doi.org/10.1080/01442872.2012.695933.

Stone, Diane (2001). "Learning Lessons, Policy Transfer and the International Diffusion of Policy Ideas." CSGR Working Paper No. 69/01. pp. 1–41.

Stromnetz Berlin online (2020). "Berlin's reliable grid operator." URL: https://www.stromnetz.berlin/en/about-us (accessed 15.09.2020).

Swan, Melanie (2015). "Blockchain. Blueprint for a New Economy." Beijing/Cambridge: O'Reilly Publishing.

Talari, Saber; Khajeh, Hosna; Shafie-khah, Miadreza; Hayes, Barry; Laaksonen, Hannu; & Catalão, João P.S. (2020). "The role of various market participants in blockchain business model." In: Shafie-khah, Miadreza (editor). Blockchain-based Smart Grids. London: Elsevier/Academic Press. pp. 75–102. DOI: https://doi.org/10.1016/B978-0-12-817862-1.00005-1.

Tapscott, Don; & Tapscott, Alex (2016). "Blockchain Revolution. How the Technology Behind Bitcoin is Changing Money, Business, and the World." New York: Penguin Random House.

Tayyar, Ari; & Gökpınar, Fatih (2019). "Green Theory in International Relations." In: Tayyar, Ari & Toprak, Elif (Editors). Theories of International Relations-II. Eskişehir: Anadolu University Publication, pp. 161–178.

Teufel, Bernd; Sentic, Anton; & Barmet, Mathias (2019). "Blockchain in future energy systems." Journal of Electronic Science and Technology. Vol 17(4). pp. 1–11. DOI: https://doi.org/10.1016/j.jnlest.2020.100011.

The White House (2023). "Building a Clean Energy Economy. A Guidebook to the Inflation Reduction Act's Investments in Clean Energy and Climate Action.." January 2023. Version 2. Washington. https://www.whitehouse.gov/wp-content/uploads/2022/12/Inflation-Reduction-Act-Guidebook.pdf. (accessed 17.05.2022).

Thomson, Camila R.; & Harrison, Gareth P. (2015). "Life cycle costs and carbon emissions of wind power: Executive Summary." ClimateXChange. The University of Edinburgh. pp. 1–23.

Tollefson, Jeff (2021). "COVID curbed carbon emissions in 2020 - but not by much." Nature. Vol. 589 (7842). URL: https://www.nature.com/articles/d41586-021-00090-3 (accessed 06.04.2023). DOI: https://doi.org/10.1038/d41586-021-00090-3.

Torbaghan, Shariat Shahab; Blaauwbroek, Niels; Kuiken, Dirk; Gibescu, Madeleine; Hajighasemi, Maryam; Nguyen, Phuong; Smit, Gerard J.M.; Roggenkamp, Martha; & Hurink, Johann (2018). "A market-based framework for demand side flexibility scheduling and dispatching." Sustainable Energy, Grids and Networks. Vol. 14. pp. 47–61. DOI: https://doi.org/10.1016/j.segan.2018.03.003.

Trading Economics (2023). "EU Carbon Permits. Trading Economics." URL: https://tradingeconomics.com/commodity/carbon (accessed 19.04.2023).

UNEP (2017). "The Emissions Gap Report 2017. A UN Environment Synthesis Report." United Nations Environment Programme (UNEP), Nairobi.

UNFCCC (no date: a). "Negotiations: Information on climate technology negotiation." URL: https://unfccc.int/ttclear/negotiations URL: (accessed 3.06.2023).

UNFCCC (no date: b). "What is the Kyoto Protocol?" URL: https://unfccc.int/kyoto_protocol (accessed 8.08.2020).

UNFCCC (no date: c) "Kyoto Protocol - Targets for the first commitment period." URL: https://unfccc.int/process-and-meetings/the-kyoto-protocol/what-is-the-kyoto-protocol/kyoto-protocol-targets-for-the-first-commitment-period (accessed 8.08.2020).

UNFCCC (no date: d). "Joint Implementation." URL: https://unfccc.int/process/the-kyoto-protocol/mechanisms/joint-implementation (accessed 8.08.2020).

UNFCCC (no date: e). "Emissions Trading." URL: https://unfccc.int/process/the-kyoto-protocol/mechanisms/emissions-trading). (accessed 8.08.2020).

UNFCCC (no date: f). "International Transaction Log." URL: https://unfccc.int/process/the-kyoto-protocol/registry-systems/international-transaction-log (accessed 8.08.2020).

UNFCCC (no date: g). "Technology Mechanism." URL: https://unfccc.int/ttclear/support/technology-mechanism.html (accessed 8.08.2020).

UNFCC (no date: h). "Parties to the United Nations Framework Convention on Climate Change." URL: https://unfccc.int/process/parties-non-party-stakeholders/parties-convention-and-observer-states?field_national_communications_target_id%5B515%5D=515&field_parties_date_of_ratifi_value=All&field_parties_date_of_signature_value=All&field_parties_date_of_ratifi_value_1=All&field_parties_date_of_signature_value_1=All&combine= (accessed 14.05.2023).

UNFCCC (2013). "Report of the Ad Hoc Working Group on the Durban Platform for Enhanced Action on the second part of its first session, held in Doha from 27 November to 7 December 2012." FCCC/ADP/2012/3 URL: https://unfccc.int/documents/7636 (accessed 16.05.2023).

UNFCCC (2008). "Report of the Conference of the Parties on its thirteenth session, held in Bali from 3 to 15 December 2007." URL: http://unfccc.int/resource/docs/2007/cop13/eng/06a01.pdf (accessed 26.05.2020).

UNFCCC (2002). "Report of the Conference of the Parties on its seventh session, held at Marrakesh from 29 October to 10 November 2001. Addendum. Part two: Action taken by the Conference of the Parties." Volume I. URL: https://unfccc.int/documents/2516 (accessed 14.05.2023).

Ungar, Michael (2011). "The Social Ecology of Resilience: Addressing Contextual and Cultural Ambiguity of a Nascent Construct." American Journal of Orthopsychiatry. Vol. 81 (1), pp. 1–17. DOI: http://dx.doi.org/10.1111/j.1939-0025.2010.01067.x.

United Nations (2019). "Climate change recognized as 'threat multiplier', UN Security Council debates its impact on peace." January 25, 2019. URL: https://www.un.org/peacebuilding/news/climate-change-recognized-'threat-multiplier'-un-security-council-debates-its-impact-peace (accessed 4.06.2020).

United Nations Secretary-General (2018). "Secretary-General's remarks on Climate Change." September 10, 2018. URL: https://www.un.org/sg/en/content/sg/statement/2018-09-10/secretary-generals-remarks-climate-change-delivered (accessed 4.06.2020).

United Nations (2016). "Transforming our World. The 2030 Agenda for Sustainable Development." A/RES/70/1. URL: https://sustainabledevelopment.un.org/content/documents/21252030%20Agenda%20for%20Sustainable%20Development%20web.pdf (accessed 20.05.2020).

United Nations General Assembly (2012). "Resolution adopted by the General Assembly. 66/290. Follow-up to paragraph 143 on human security of the 2005 World Summit Outcome." A/RES/66/290. 25 October 2012. URL: https://www.unocha.org/sites/dms/HSU/Publications%20and%20Products/GA%20Resolutions%20and%20Debate%20Summaries/GA%20Resolutions.pdf (accessed 5.06.2023).

United Nations Treaty Collection (2023). "Chapter XXVII 7. Environment." URL: https://treaties.un.org/Pages/ViewDetailsIII.aspx?src=IND&mtdsg_no=XXVII-7&chapter=27&Temp=mtdsg3&clang=_en (accessed 14.05.2023).

Urban, Frauke (2018). "China's rise: Challenging the North-South technology transfer paradigm for climate change mitigation and low carbon energy." Energy Policy. Vol. 113. pp. 320–330. DOI: 10.1016/j.enpol.2017.11.007.

Urrutia Silva, Osvaldo (2010). "El régimen jurídico internacional del cambio climático después del 'Acuerdo de Copenhague.'" Revista de Derecho de la Pontificia Universidad Católica de Valparaíso [online]. Vol. 34, pp. 597–633. DOI: 10.4067/S0718-68512010000100019.

US Department of Energy (2013). "Comparing the Impacts of Northeast Hurricanes on Energy Infrastructure." Office of Electricity Delivery and Energy Reliability. pp1–44. URL: https://www.oe.netl.doe.gov/docs/Northeast%20Storm%20Comparison_FINAL_041513c.pdf.

von Arnim, Achaz; & von Arnim, Julius (2020). "E-Mobility 4.0 - erfolgreiches Zusammenspiel von Prosumern mit Energieeffizienzhäusern und Stadtwerken." In: Doleski, Oliver D. (Editor). Realisierung Utility 4.0, Vol. 2. Praxis der digitalen Energiewirtschaft vom Vertrieb bis zu innovativen Energy Services. Wiesbaden: Springer Vieweg, pp. 797–814. DOI: 10.1007/978-3-658-25589-3_53.

Wang, Shuai; Ouyang, Liwei; Yuan, Yong; Ni, Xiaochun; Han, Xuan; & Wang, Fei-Yue (2019). "Blockchain-Enabled Smart Contracts: Architecture, Applications, and Future Trends." IEEE Transactions on Systems, Man, and Cybernetics: Systems. Vol. 49(11). pp. 2266–2277. DOI: 10.1109/TSMC.2019.2895123.

Wang, Jian; Wang, Qianggang; Zhou, Niancheng; & Chi, Yuan (2017). "A Novel Electricity Transaction Mode of Microgrids Based on Blockchain and Continuous Double Auction." Energies. Vol. 10 (12), pp. 1–22. DOI: 10.3390/en10121971.

Wei, Jiapeng; Wulan, Bulbul; Yan, Jiaqi; Sun, Mengjia; & Jing, Hong (2019). The Adoption of Blockchain Technologies in Data Sharing: A State of the Art Survey". WHICEB 2019. Proceedings. Vol. 59. pp. 54–61. URL: https://aisel.aisnet.org/whiceb2019/59

Wilkins, Gill (2002). "Technology Transfer for Renewable Energy. Overcoming Barriers in Developing Countries." London: Earthscan Publications.

WindNODE (no date: a). "Funding and Political Patronage." https://www.windnode.de/en/about/funding-and-political-patronage/overview/ (accessed 10.09.2020).

WindNODE (no date: b). "WEMAG Netz GmbH." https://www.windnode.de/en/partners/wemag-netz-gmbh/ (accessed 15.09.2020).

WindNODE (no date: c). "ENSO Netz GmbH." https://www.windnode.de/en/partners/associated-partners/enso-netz-gmbh/ (accessed 15.09.2020).

WindNODE (no date: d). "E.DIS. Netz GmbH." https://www.windnode.de/en/partners/associated-partners/edis-netz-gmbh/ (accessed 15.09.2020).

WindNODE (2020). "Das Schaufenster für intelligente Energie aus dem Nordosten Deutschland 2017–2020." URL: https://www.windnode.de/fileadmin/Daten/Downloads/Jahrbuch/WindNODE_Jahrbuch_2020_Web_150dpi.pdf (accessed 2.06.2023).

WindNODE (2019). "Flexibility Platform: First Ever Capacity Offers and Calls." 03.15.2019. https://www.windnode.de/en/newsdetail/news/flexibilitaetsplattform-erstmals-kapazitaeten-angeboten-und-abgerufen/?tx_news_pi1%5Bcontroller%5D=News&tx_news_pi1%5Baction%5D=detail&cHash=73276efb12de9215ce9e1085125a4728 (accessed 11.09.2020).

WindNODE (2018). "Utilization before Limitation." 20.11.2018. URL: https://www.windnode.de/en/newsdetail/news/nutzen-statt-abregeln/?tx_news_pi1%5Bcontroller%5D=News&tx_news_pi1%5Baction%5D=detail&cHash=b6f42162245dc1cf24b4a609dd222021 (accessed 11.9.2020).

World Bank (2022). "Digital Monitoring, Reporting, and Verification Systems and Their Application in Future Carbon Markets." World Bank, Washington, DC.

World Bank (2020). "Tracking SDG 7. The Energy Progress Report 2020." URL: https://trackingsdg7.esmap.org/data/files/download-documents/tracking_sdg_7_2020-full_report_-_web_0.pdf (accessed 25.09.2020).

World Bank (2019). "Tracking SDG 7. The Energy Progress Report 2019." URL: https://www.worldbank.org/en/topic/energy/publication/tracking-sdg7-the-energy-progress-report-2019. (accessed 25.09.2020).

World Bank (2018). "Blockchain and Emerging Digital Technologies for Enhancing Post-2020 Climate Markets." The World Bank Group Climate Change, pp.1–28.

Worldcoin (22.03.2023). "What's Ethereum 2.0? A Complete Guide." URL: https://worldcoin.org/articles/whats-ethereum20#:~:text=In%20terms%20of%20processing%20speed,around%2030%20transactions%20per%20second. (accessed 4.05.2023).

van Ruijven, Bas J.; De Cian, Enrica, & Sue Wing, Ian (2019). "Amplification of future energy demand growth due to climate change." Nature Communication. Vol. 10, 2762 (2019). p. 1–12. DOI: 10.1038/s41467-019-10399-3.

Vivekananda, Janani (2022). "Reimagining the Human-Environmental Relationship." United Nations University. Centre for Policy Research. URL: https://collections.unu.edu/eserv/UNU:8836/UNUUNEP_Vivekananda_RHER.pdf.

Vogler, John (2018). "Energy, Climate Change, and Global Governance: The 2015 Paris Agreement in Perspective." In: Davidson, Debra J.; & Gross, Matthias (editors). Oxford Handbook of Energy and Society. Oxford: Oxford University Press, pp. 15–30. DOI: 10.1093/oxfordhb/9780190633851.013.0002.

Xu, Yueqiang; Ahokangas, Petri; Louis, Jean-Niclas, & Pongrácz, Eva (2019). "Electricity Market Empowered by Artificial Intelligence: A Platform Approach." Energies, Vol. 12(21), pp. 1–21. DOI: 10.3390/en12214128.

Yergin, Daniel (2006). "Ensuring Energy Security." Foreign Affairs. Vol. 85(2). pp. 69–82. DOI: 10.2307/20031912.

Yoldas, Yeliz; Önen, Ahmet, Muyeen, S. M., Vasilakos, Athanasios Vasilakos; & Alan, Irfan (2017). "Enhancing smart grid with microgrids: Challenges and opportunities." Renewable and Sustainable Energy Reviews, Vol. 72, pp. 205–214. DOI: 10.1016/j.rser.2017.01.064.

Yoshida, Fumikazu (2012). "The Theory of Environment Governance." (Chapter 4). In: Lecture on Environmental Economics. Hokkaido University. pp. 75–103. URL: https://eprints.lib.hokudai.ac.jp/dspace/bitstream/2115/53454/1/chapter-4.pdf. (accessed 22.11.2022).

Young, Oran R. (2005). "Why is there no unified theory of environmental governance?" In: Dauvergne, Peter (Editor). Handbook of Global Environmental Politics. Cheltenham/Northampton: Editorial Edward Elgar. pp. 170–184.

Zaman, Sharaban Tahura (2018). "The 'bottom-up pledge and review' approach of nationally determined contributions (NDCs) in the Paris Agreement: A Historical Breakthrough or a setback in new climate governance?" IALS Student Law Review, Vol. 5 (2). pp. 3–20. DOI: 10.14296/islr.v5i2.4898.

Zheng, Zibin; Xie, Shaoan; Chen, Xiangping; & Wang, Huaimin (2018). "Blockchain challenges and opportunities: a survey." International Journal of Web and Grid Services, Vol. 14 (4), p. 352–375. DOI:10.1504/IJWGS.2018.095647.

Zhou, Qiheng.; Huang, Huawei; Zheng, Zibin; & Bian, Jiing (2020). "Solutions to Scalability of Blockchain: A Survey." IEEE Access. Vol. 8, pp. 16440–16455. DOI: 10.1109/ACCESS.2020.2967218.

Zia, Muhammad F.; Benbouzid, Mohamed; Elbouchikhi, Elhoussin; Muyeen, S. M., Techato, Kuaanan; Guerrero; & Guerrero, Josep M. (2020). "Microgrid Transactive Energy—Review, Architectures, Distributed Ledger Technologies, and Market Analysis." IEEE Access, pp. 19410–19432. DOI: 10.1109/ACCESS.2020.2968402.

50Hertz online (no date). "This is 50Hertz." URL: https://www.50hertz.com/en/Company (accessed 15.09.2020).

50Hertz & WindNODE (2018). "Nutzen statt Abregeln. Die WindNODE Flexibilitätsplattform." URL: https://www.google.com/url?sa=t&rct=j&q=&esrc=s&source=web&cd=&ved=2ahUKEwiO2Ne7uaX_AhXPbPEDHSn-ABcQFnoECA4QAQ&url=https%3A%2F%2Fwww.50hertz.com%2FPortals%2F1%2FDokumente%2FMedien%2Fpublikationen%2FBrochure%2520Flex%2520plattform.pdf&usg=AOvVaw29nHhKlho-OdtZleWTE9WY (accessed, 2.06.2023).

10.1. Interviews:

Interview conducted with the technical lead of blockchain prototype Philipp Lämmel at Fraunhofer FOKUS on July 28th, 2020, Phillip Lämmel (WindNODE) on July 28th, 2020 and September 29th, 2020.

Interview conducted with Juan Ignacio Ibañez (Centre for Blockchain Technologies, University College London) on September 29th, 2020.